Lecture Notes in Computer Science 13082

Claus Aranha · Carlos Martín-Vide ·
Miguel A. Vega-Rodríguez (Eds.)

Theory and Practice of Natural Computing

10th International Conference, TPNC 2021
Tsukuba, Japan, December 7–10, 2021
Proceedings

 Springer

Editors
Claus Aranha ⓘ
Graduate School of Systems
and Information Engineering
University of Tsukuba
Tsukuba, Japan

Carlos Martín-Vide ⓘ
Research Group on Mathematical
Linguistics
Rovira i Virgili University
Tarragona, Spain

Miguel A. Vega-Rodríguez ⓘ
School of Technology
University of Extremadura
Cáceres, Spain

ISSN 0302-9743   ISSN 1611-3349   (electronic)
Lecture Notes in Computer Science
ISBN 978-3-030-90424-1   ISBN 978-3-030-90425-8   (eBook)
https://doi.org/10.1007/978-3-030-90425-8

LNCS Sublibrary: SL1 – Theoretical Computer Science and General Issues

This Springer imprint is published by the registered company Springer Nature Switzerland AG
The registered company address is: Gewerbestrasse 11, 6330 Cham, Switzerland

Preface

These proceedings contain the papers that were presented at the 10th International Conference on the Theory and Practice of Natural Computing (TPNC 2021), held in Tsukuba, Japan, during December 7–10, 2021.

The scope of TPNC is rather broad, including the following:

- Theoretical contributions to affective computing, ambient intelligence, ant colony optimization, approximate reasoning, artificial immune systems, artificial life, cellular automata, cognitive computing, cognitive robotics, collective intelligence, combinatorial optimization, computational intelligence, computing with words, developmental systems, DNA computing, evolutionary algorithms, evolutionary computing, evolutionary game theory, fuzzy logic, fuzzy sets, fuzzy systems, genetic algorithms, genetic programming, global optimization, granular computing, heuristics, intelligent agents, intelligent control, intelligent manufacturing, intelligent systems, intelligent user interfaces, machine intelligence, membrane computing, metaheuristics, molecular programming, multi-objective optimization, neural networks, quantum communication, quantum computing, quantum information, quantum metrology, rough sets, soft computing, swarm intelligence, swarm robotics, and unconventional computing.
- Applications of natural computing to algorithmics, bioinformatics, cryptography, design, economics, graphics, hardware, human-computer interaction, knowledge discovery, learning, logistics, medicine, natural language processing, pattern recognition, planning and scheduling, programming, telecommunications, and web intelligence.

TPNC 2021 received 14 submissions, and the papers were reviewed by three Program Committee members. There were also a few external reviewers consulted. After a thorough and vivid discussion phase, the committee decided to accept nine papers (which represents an acceptance rate of 64%). The conference program also included three invited talks as well as some presentations of work in progress.

The excellent facilities provided by the EasyChair conference management system allowed us to deal with the submissions successfully and handle the preparation of these proceedings in time.

We would like to thank all invited speakers and authors for their contributions, the Program Committee and the external reviewers for their cooperation, and Springer for its very professional publishing work.

September 2021

Claus Aranha
Carlos Martín-Vide
Miguel A. Vega-Rodríguez

Organization

TPNC 2021 was organized by the University of Tsukuba, Japan, and the Institute for Research Development, Training and Advice (IRDTA), from Brussels, Belgium, and London, UK.

Program Committee

Andrew Adamatzky	University of the West of England, UK
Claus Aranha	University of Tsukuba, Japan
Peter J. Bentley	University College London, UK
Erik Cambria	Nanyang Technological University, Singapore
Christer Carlsson	Åbo Akademi University, Finland
Shyi-Ming Chen	National Taiwan University of Science and Technology, Taiwan
Claude Crépeau	McGill University, Canada
Ernesto Damiani	University of Milan, Italy
Yong Deng	University of Electronic Science and Technology of China, China
Matthias Ehrgott	Lancaster University, UK
Étienne Kerre	Ghent University, Belgium
Sam Kwong	City University of Hong Kong, Hong Kong
Chung-Sheng Li	PricewaterhouseCoopers, USA
Jing Liang	Zhengzhou University, China
Robert Mann	University of Waterloo, Canada
Carlos Martín-Vide (Chair)	Rovira i Virgili University, Spain
Luis Martínez López	University of Jaén, Spain
Serge Massar	Université Libre de Bruxelles, Belgium
Marjan Mernik	University of Maribor, Slovenia
Seyedali Mirjalili	Torrens University, Australia
Ngoc Thanh Nguyen	Wrocław University of Science and Technology, Poland
Leandro Nunes de Castro	Mackenzie Presbyterian University, Brazil
Matjaž Perc	University of Maribor, Slovenia
Brian M. Sadler	Army Research Laboratory, USA
Patrick Siarry	Paris-Est Créteil University, France
Andrzej Skowron	University of Warsaw, Poland
Stephen Smith	University of York, UK
Ponnuthurai N. Suganthan	Nanyang Technological University, Singapore
Vicenç Torra	Umeå University, Sweden
Rufin VanRullen	CNRS Toulouse, France
Miin-Shen Yang	Chung Yuan Christian University, Taiwan
Yi Zhang	Sichuan University, China

Organizing Committee

Claus Aranha (Co-chair)	University of Tsukuba, Japan
Yuri Lavinas	University of Tsukuba, Japan
Sara Morales	IRDTA, Belgium
Manuel Parra-Royón	University of Granada, Spain
David Silva (Co-chair)	IRDTA, UK
Miguel A. Vega-Rodríguez	University of Extremadura, Spain

Additional Reviewer

Kaushik Das Sharma

Contents

Applications of Natural Computing

Computing the Optimal Longest Queue Length in Torus Networks

Mehrdad Aliasgari[1] , Burkhard Englert[2], and Oscar Morales-Ponce[1]([⊠])

[1] Department of Computer Engineering and Computer Science,
California State University Long Beach, Long Beach, USA
{mehrdad.aliasgari,oscar.morales-ponce}@csulb.edu
[2] Department of Computer Science, University of North Caroline Wilmington,
Wilmington, USA

Abstract. A collection of k mobile agents is arbitrarily deployed in the edges of a directed torus network where agents perpetually move to the successor edge. Each node has a switch that allows one agent of the two incoming edges to pass to its successor edge in every round. The goal is to obtain a switch scheduling to reach and maintain a configuration where the longest queue length is minimum. We consider a synchronous system. We use the concept of conflict graphs to model the local conflicts that occur with incident links. We show that there does not exist an algorithm that can reduce the number of agents in any conflict cycle of the conflict graph providing that all the links have at least 2 agents at every round. Hence, the lower bound is at least the average queue length of the conflict cycle with the maximum average queue length. Next, we present a centralized algorithm that computes a strategy in $O(n \log n)$ time for each round that attains the optimal queue length in $O(\sigma n)$ rounds where n is the number of nodes in the network and σ is the standard deviations of the queue lengths in the initial setting. Our technique is based on network flooding on conflict graphs. Next, we consider a distributed system where nodes have access to the length of their queues and use communication to self-coordinate with nearby nodes. We present a local algorithm using only the information of the queue lengths at distance two. We show that the algorithm attains the optimal queue length in $O(\sigma C_{max}^2)$ rounds where C_{max} is the length of the longest conflict cycle with the maximum average queue length.

Keywords: Mobile agents · Torus networks · Lower bound

1 Introduction

Protecting a region with a collection of agents has recently attracted the attention of the distributed computing community. The goal of theses problems is to understand the interaction of the agents, provide lower bounds that any algorithm can attain as well as providing strategies that guarantee optimal or approximation upper bounds. In this paper, we study a variant of the problem where

© Springer Nature Switzerland AG 2021
C. C. Aranha et al. (Eds.): TPNC 2021, LNCS 13082, pp. 3–14, 2021.
https://doi.org/10.1007/978-3-030-90425-8_1

agents are arbitrarily deployed in a graph. Agents perpetually move along a given cycle. However, agents on different cycles may collide after crossing conflicting nodes. To avoid collisions, a switch at each node coordinates the agents. The goal is to reach and maintain a configuration where the longest queue length is minimum. However, the use of switches can result in unbalancing the number of agents in each link. Is it possible to provide a switch schedule that guarantees an even distribution of the agents? To our knowledge, there are no formal studies that provide lower and tight upper bounds. In this paper, we address a deterministic version of the problem in symmetric networks. Specifically, we study the problem in torus networks of dimension $\sqrt{n} \times \sqrt{n}$, i.e., the torus has $\sqrt{n}$ horizontal and $\sqrt{n}$ vertical cycles or rings. Initially, a collection of k agents is arbitrarily deployed in the rings such that each ring has $k\sqrt{n}$ agents. To simplify the presentation, we refer to the links as queues to denote the order of the agents. We consider a synchronous system where each node is crossed by two agents in each round. The natural question is if it is always possible to reach a deployment where all the links have the same number of agents, i.e., queue length k. We can answer the question by determining the minimum longest queue length in each link that can be reached. To reduce the queue length, the agent in front of the queue can cross to the successor link meanwhile, the agent in front of the predecessor queue does not cross. However, this may increase the queue length of its incident links. We call the links that are negatively affected by the scheduling conflict links. Further, the conflicts can extend into conflict paths, which in turn can form conflict cycles. We model these conflicts using *conflict graphs*.

We show that there does not exist an algorithm that can reduce the number of agents in any conflict cycle if the links have at least 2 agents at every round. Hence, the optimal lower bound is at least the maximum average length among all the conflict cycles in the network. Next, we present a centralized algorithm for directed torus networks that computes a strategy in $O(n \log n)$ time that minimizes the longest queue length in $O(\sigma n)$ rounds where n is the size of the network and σ is the standard deviation of the queue lengths in the initial deployment. The algorithm uses the conflict graph to apply a technique called flooding, used in routing algorithms, to minimize the longest queue length. Next, we present a local algorithm that uses the queue length of links at a distance two to minimize the longest queue length in at most $O(\sigma C_{max}^2)$ rounds where C_{max} is the length of the longest conflict cycle with maximum average queue length in the initial deployment. We observe that our results can be applied to non-deterministic settings and general topologies.

Organization of the Paper. We present the formal model and the problem statement in Sect. 1.1. The related work is presented in Sect. 2. The concept of conflict graph is presented in Sect. 3 and the lower bound in Sect. 4. In Sect. 5, we present the centralized algorithm meanwhile, the local algorithm is presented in Sect. 6. We conclude the paper in Sect. 7. A complete version of the paper can be found on https://arxiv.org/abs/1606.03800.

1.1 Model

Let $G = (V, E)$ denote a graph where V is the set of vertices (or nodes) and $E = \{\{u, v\} : u, v \in V\}$ is the set of links. An orientation of G is the directed graph obtained by assigning to each link a direction. Let $(u, v) \in \overrightarrow{E}$ be the directed link with source at u and destination at v. We consider a torus graph of dimension $m \times m$ which consists of m horizontal and m vertical rings. More specifically, a torus network of dimension $m \times m$ is the graph $G = (V, E)$ where $V = \{v_{i,j} | \forall i \in [0, m-1] \wedge \forall j \in [0, m-1]\}$ is the set of vertices and $E = \{\{v_{i,j}, v_{k,l}\} | (k = i + 1 \bmod m \wedge l = j) \vee (k = i \wedge l = j + 1 \bmod m)\}$ is the set of links. Thus, the number of vertices is $n = m^2$ and the number of links is $2n$. We assume that each link has a queue where agents wait. Throughout the paper, we refer to the ring formed with links $\{v_{i,j}, v_{i,(j+1) \bmod m}\}$ for all $j \in [0, m-1]$ as the horizontal ring i. Similarly, we refer to the ring formed with links $\{v_{i,j}, v_{(i+1) \bmod m,j}\}$ for all $i \in [0, m-1]$ as the vertical ring j. Observe that each vertex $v_{i,j}$ has two incoming and two outgoing links. For each v, let $\overrightarrow{E}^{in}(v)$ denote the set of incoming links to v and $\overrightarrow{E}^{out}(v)$ denote the set of outgoing links from v. Given a link $e = (u, v)$, let $succ(e), pred(e)$ denote the successor and predecessor link of e as defined by the direction of the ring connecting u and v. To simplify our presentation, we consider m to be an even number and a torus where the direction of the rings alternate. However, our solutions and analysis also work in tori where the rings have any arbitrarily direction.

Let Ω be the set of agents arbitrarily deployed in the link queues of the torus. We divide the time in rounds and in every round, a semaphore in each vertex v allows two agents crossing to their successor links, i.e., let $e, e' \in \overrightarrow{E}^{in}(v)$, the semaphore either allows an agent in e and e' moving to $succ(e)$ and $succ(e')$, respectively, or two agents in e moving to $succ(e)$. Observe that each agent infinitely often traverses its initial ring. Therefore, the number of agents in each ring remains the same at all times. We refer to the crossing schedule in each vertex as the green time assignment. Let $g_e(r)$ denote the green time of e at round r. The green time assignment has the constraint that $g_e(r) + g_{e'}(r) = 2$ where e and e' are in $\overrightarrow{E}^{in}(v)$. There are two ways to reduce the queue length of $e = (u, v)$ in one round:

1. Allowing two agents in front of e moving to $succ(e)$. However, the queue length at the other incoming link incident to v, i.e., $(u', v) = orth(e)$ as well as the queue length of $succ(e)$ increase. Thus, e is in forward conflict with $succ(e)$ and $orth(e)$.
2. Preventing the agents in front of $pred(e)$ to move to e. However, the queue length of the other outgoing link from u, i.e., $(u, w) = borth(e)$ as well as the queue length of $pred(e)$ increase. Thus, e is in backward conflict with $pred(e)$ and $borth(e)$.

Let $w_e(r)$ be the number of agents in e at round r. We define the agents deployment at time r as $W(r) = \{w_e(r) : \forall e \in E\}$. Throughout the paper, we usually omit the time reference if it is clear from the context.

Definition 1. *Given an agent deployment* $W(r) = \{w_e(r) : \forall e \in E\}$ *and a set* $A(r) = \{g_e(r) : \forall e \in E\}$ *of green time assignments, the deployment of* $e \in E$ *at round* $r + 1$ *is given by:*

$$w_e(r+1) := w_e(r) - \min(g_e(r), w_e(r)) + \min(g_{pred(e)}(r), w_{pred(e)}(r)).$$

We say that the network is saturated if $w_e(r) \geq 2$ for all $e \in E$. Therefore, we can simplify the model in saturated torus networks and rewrite it as:

$$w_e(r+1) := w_e(r) - g_e(r) + g_{pred(e)}(r).$$

Problem Statement: Given an initial agent deployment W in an oriented torus $G = (V, E)$, the overall objective is to determine the green time assignments such that the longest queue is minimum after a finite number of rounds. Let $\mathcal{A}(r)$ denote the set of green time assignments at round r. Formally,

Problem: Determine the green time assignments $\mathcal{A}(0), \mathcal{A}(1), ..., \mathcal{A}(r), ...$ such that there exists a time r when the longest queue length overall $e \in E$ is minimum thereafter. Let ϕ denote the minimum longest queue length.

2   Related Work

Kortsarz and *Peleg* [6] studied the traffic-light scheduling for route scheduling on a two-dimensional grid. They show upper bounds according to the number of directions. Unlike our model where each vertex always allows one agent crossing (in either direction), edges can be activated and deactivated at any round. Moreover, in our setting agents infinitely often traverse the grid in one direction. Our problem is closely related to the packet switching networks where switches can forward one packet at the time. The authors in [1] propose isarithmic networks where a constant number of agents (packets) in the overall network is maintained. When a packet arrives at its destination, it is replaced with new payload before putting back into the system. In our model, the number of agents remains constant at every time. Agents can be used to transmit data. Therefore, our model can be considered an isarithmic network and our results can be applied to these networks.

Our problem is also related to the routing problem in high-performance computers where cores are interconnected with different topologies such as torus networks. Cores then communicate sending messages which are routed through the network. Its performance and efficiency largely depend on routing techniques. Many authors have proposed different routing techniques that deal with the throughput, e.g., [2–5]. Despite providing optimal routes with respect with some metric, it is not clear how concurrent routes affect the system. In other words, how the system capacity is affected with a high volume of concurrent routes. Although we do not deal directly with routing, our problem provides the minimum longest queue that can be used as a primary metric for the system.

3 Traffic Network Flows

In this section, we introduce the concept of the conflict graph of an oriented torus $G = (V, \overrightarrow{E})$ that allows us applying flooding techniques to G in a natural way. Next, we introduce the green time shifts that we use to define the shift assignments. Finally, we introduce conflict cycles in the conflict graphs.

Intuitively, in the conflict graph, the links of the torus are the set of vertices and two vertices u, v are adjacent if reducing the queue length of u in the direction of v increases the queue length of v.

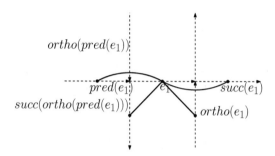

Fig. 1. Conflict links (Dotted lines denote the links of the torus and solid lines the links of the conflict the conflict graph.)

Definition 2 (Conflict Graph). *Given an oriented torus $G = (V, \overrightarrow{E})$ the conflict graph of G, is $L(G) = (E, \overrightarrow{E}')$; see Fig. 1, where for any $e_1, e_2 \in E$ there exists an edge $(e_1, e_2) \in \overrightarrow{E}'$ if and only if one of the following conditions holds:*

1. *$e_2 = succ(e_1)$ (Equivalent $e_1 = pred(e_2)$).*
2. *$\exists v : \{e_1, e_2\} \subseteq \overrightarrow{E}^{in}(v)$; i.e., $e_2 = orth(e_1)$.*
3. *$\exists v : \{e_1, e_2\} \subseteq \overrightarrow{E}^{out}(v)$; i.e., $e_2 = borth(e_1)$).*

The definition of conflict graphs can be extended to other digraphs. For a fixed link $e \in \overrightarrow{E}$, we denote the set of forward conflict neighbors of e as $N^+(e) = \{succ(e), orth(e)\}$ and the set of backward conflict neighbors as $N^-(e) = \{pred(e), borth(e)\}$.

For each link e, let $g_e = [0, 2]$ be the green time shift subject to

$$g_e + g_{orth(e)} = 2. \tag{1}$$

Let $A(r)$ be the set of green time shifts.

Definition 3 (Traffic Network). *Let $L(G)$ be the conflict graph of $G = (V, \overrightarrow{E})$ with an agent deployment $W(r)$ at round r. A network is defined by the tuple $TN(r) = (L(G), W(r), A(r))$. At the beginning of every round, $s_e(r) = 0$ for all $e \in \overrightarrow{E}$.*

Throughout the paper, we refer to the network as $TN = (L(G), W, A)$ if it is clear from the context.

For every link $e = (u, v) \in \overrightarrow{E}$, we can reduce its queue length by setting $g_e = 2$. However, from Constraint 1, $g_{orth(e)} = 0$. We call this operation forward flow. Similarly, we can reduce its queue length by setting $g_{pred(e)} = 0$. However, from Constraint 1, $g_{borth(e)} = 2$. We call this operation backward flow. The following definition formalizes the flows.

Definition 4 (Modified flows in Traffic Network). *The forward flow change operation f_e^+ sets $g_e = 2$ and $g_{orth(e)} = 0$. Analogously, the backward flow change operation f_e^- sets $g_{pred(e)} = 0$ and $g_{orth(pred(e))} = 2$.*

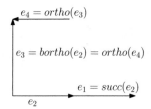

Fig. 2. A conflict path.

Next, we introduce the concept of conflict path P in TN; refer to Fig. 2. We say that a path $P = \{e_1, e_2, ..., e_l\}$ in TN is a conflict path if either $|P| = 2$ and e_1 and e_2 are in conflict or $|P| \geq 2$ and for every three consecutive links $e_j, e_{j+1}, e_{j+2} \in P$, exactly one of the following statement follows:

1. if $e_{j+1} = orth(e_j)$ or $e_{j+1} = pred(e_j)$,
   then $e_{j+2} \in N^-(e_{j+1})$
2. if $e_{j+1} = succ(e_j)$ or $e_{j+1} = borth(e_j)$,
   then $e_{j+2} \in N^+(e_{j+1})$

We define a conflict cycle C of length l to be a closed conflict path, i.e., $C = \{e_0, e_1, ..., e_{l-1}, e_0\}$; see Fig. 3.

We say that a link $e \in \overrightarrow{E}$ is an *exit link* if e is in the conflict cycle C but $succ(e)$ is not in C. Similarly, a link $e \in \overrightarrow{E}$ is an *entry link* if e is not in C, but $succ(e)$ is in C. A vertex $v \in C$ is an *exit cycle* if both links in $\overrightarrow{E}^{out}(v)$ are exit links. A vertex $v \in C$ is an *entry cycle* if both links in $\overrightarrow{E}^{in}(v)$ are entry links. Let $V^-(C)$ and $V^+(C)$ denote the set of exit cycles and the set of entry cycles in C, respectively.

Lemma 5. *In any conflict cycle C, the number of entry vertices is equal to the number of exit vertices, i.e., $|V^+(C)| = |V^-(C)|$.*

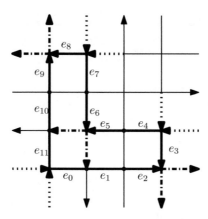

Fig. 3. A conflict cycle is represented with bold solid links, dotted arrows are entry links and dashed dotted arrows are exit links.

4   Lower Bound

In this section, we present the lower bound on the minimum longest queue that any algorithm can attain in saturated networks. Recall that a saturated network is a TN where $w_e(r) \geq 3$ for each $e \in TN$.

Observe that a conflict cycle is, indeed, equivalent to a deadlock in the sense that reducing the number of agents in any queue of a cycle increases the queue length of another link in the cycle.

Lemma 6. *Let* $TN = (L(G), W, A)$ *be a saturated network and let* $\mathcal{C}$ *be the set of all conflict cycles in* $L(G)$. *Then, the number of agents in each conflict cycle* $C \in \mathcal{C}$ *remains constant for all rounds.*

From Lemma 6, the best that any algorithm can do is to uniformly distribute the agents along the conflict cycle with the maximum average of agents as shown in the following theorem.

Theorem 7. *Let* $TN = (L(G), W, S)$ *be a saturated network. Then the longest queue length* ϕ *that any greedy algorithm can attain is at least*

$$\phi \geq \max_{C \in \mathcal{C}} \left(\left\lceil \frac{\sum_{e \in C} w_e(r)}{|C|} \right\rceil \right)$$

where $\mathcal{C}$ *is the set of all conflict cycles in* $L(G)$.

In the sequel, we say that C is a critical conflict cycle if $\phi = \left\lceil \frac{\sum_{e \in C} w_e}{|C|} \right\rceil$.

5   Minimizing the Longest Queue Length in Saturated Traffic Networks

In this section, we present a strategy that minimizes the maximum queue length in saturated networks. The main idea of the algorithm is based on flooding in distributed computing using Definition 4. We start the flooding in a queue with the longest queue length ϕ' which provokes that the conflict links get flooded. We continue flooding the conflict links until all the queue lengths in the flooding paths are less than ϕ'.

Given a network $TN = (L(G), W, A)$, an i-conflict path $e_1, e_2, ..., e_a$ of TN is a conflict path where $w_{e_1} = i$ and $w_{e_j} = i - 1$ for all $j \in [2, a]$. An i-conflict path $e = e_1, e_2, ..., e_a$ is *forward i-conflict path* if $e_2 \in N^+(e)$. Otherwise, is a *backward i-conflict path*. We say that e' is the *forward i-conflict adjacent* of e if there exists a conflict path $e = e_1, e_2, .., e_a = e'$ such that $w_{e_1} = w_{e_a} = i$ and $w_{e_j} = i - 1$ for $j \in [2, a - 1]$; see Fig. 4. Analogously, we define the *backward i-conflict adjacent*. Observe that it is always possible to reduce the queue length by flooding the i-conflict path. On the other hand, the path between two links that are i-conflict adjacent cannot be reduced.

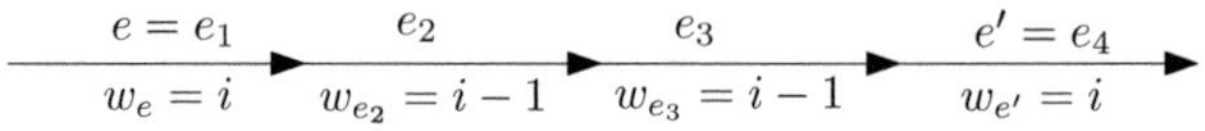

Fig. 4. e' is the forward i-conflict adjacent of e and e is the backward i-conflict adjacent of e'

Consider a link e, the forward indicative function returns false if e does not have a forward i-conflict adjacent neighbor and true if there exists e' such e' is the forward i-conflict adjacent of e. Formally,

$$\mathbb{1}_e^+ = \begin{cases} true & \text{if } \exists e' \text{ such that } e \text{ and } e' \text{ are forward } i\text{-conflict adjacent} \\ false & \text{otherwise} \end{cases}$$

Similarly we define for backward i-conflict adjacent, i.e., $\mathbb{1}_e^-$.

Our approach consists of two steps. First, we present an efficient algorithm to find a link with maximum queue length such that the value of either $\mathbb{1}_e^+$ or $\mathbb{1}_e^-$ is *false*. Once we identify a link e such that either $\mathbb{1}_e^+$ or $\mathbb{1}_e^-$ is *false* we reduce its queue length. These two steps are repeated until there is no link e with maximum queue length such that neither $\mathbb{1}_e^+$ nor $\mathbb{1}_e^-$ is *false*.

A conflict tree $T_e^+(r)$ is a tree of forward i-conflict paths without cycles at round r where $w_e = i$ and $w_{e'} = w_e - 1$ for each $e' \in T_e^+(r)$ different from e. Similar, we define $T_e^-(r)$. $T_e^{dir}(r)$ is maximal if it cannot be extended in the direction dir.

Next, we present an algorithm that given a link e with the maximum queue length it determines if either $\mathbb{1}_e^+$ or $\mathbb{1}_e^-$ is *false*. Observe that the naive algorithm for determining the values $\mathbb{1}_e^+$ and $\mathbb{1}_e^-$ takes quadratic time as every conflict path is traversed and potentially there exist a quadratic number of conflict paths. Instead, we construct $T_e^{dir}(r)$ until $T_e^+(r)$ is maximal or a conflict cycle appears.

Lemma 8. *Given e and dir such that w_e is the largest queue length and $dir \in \{+, -\}$, there exists a linear time algorithm that determines whether $\mathbb{1}_e^+ = false$ or e is part of an i-conflict cycle.*

Algorithm 1: AdjacentConflict:

Input: $root$: such that the queue length is maximum

Input: dir: $\{+, -\}$

1 $i \leftarrow w_{root}$;

2 Let Q be a stack;

3 $push(Q, (root, dir))$;

4 **while** $Q \neq \emptyset$ **do**

5 $\quad$ $(e, dir) \leftarrow pop(Q)$;

$\quad$ /* A link is red if it has breen visted   */

6 $\quad$ **if** $color((e, dir)) = red$ **then**

7 $\quad\quad$ **return** There is a w_{root}-conflict cycle;

8 $\quad$ $color((e, dir)) = red$;

9 $\quad$ **if** $dir = +$ **then**

10 $\quad\quad$ **if** $w_{succ(e)} = i$ **then**

11 $\quad\quad\quad$ $\mathbb{1}^{-}_{succ(e)} = true$;

12 $\quad\quad$ **else if** $w_{succ(e)} = i - 1$ **then**

13 $\quad\quad\quad$ $push(Q, (succ(e), +))$

14 $\quad\quad$ **if** $w_{orth(e)} = i$ **then**

15 $\quad\quad\quad$ $\mathbb{1}^{+}_{orth(e)} = true$;

16 $\quad\quad$ **else if** $w_{orth(e)} = i - 1$ **then**

17 $\quad\quad\quad$ $color(Q, (orth(e), -))$

18 $\quad$ **else**

19 $\quad\quad$ **if** $w_{pred(e)} = i$ **then**

20 $\quad\quad\quad$ $\mathbb{1}^{+}_{orth(e)} = true$;

21 $\quad\quad$ **else if** $w_{pred(e)} = i - 1$ **then**

22 $\quad\quad\quad$ $push(Q, (pred(e), -))$

23 $\quad\quad$ **if** $w_{borth(e)} = i$ **then**

24 $\quad\quad\quad$ $\mathbb{1}^{-}_{borth(e)} = true$;

25 $\quad\quad$ **else if** $w_{borth(e)} = i - 1$ **then**

26 $\quad\quad\quad$ $push(Q, (borth(e), +))$

27 **return** $false$;

Algorithm 1 can be used to find the value of $\mathbb{1}^{dir}_{e}$ for every link with the longest queue length.

Lemma 9. *Given a network, we can compute $\mathbb{1}^{+}_{e}$ and $\mathbb{1}^{-}_{e}$ for each link e with the maximum queue length in linear time.*

We show in the next lemma how the queue length of a link e with maximum queue length such $\mathbb{1}^{-}_{e} = false$ can be reduced. The main observation is that when $\mathbb{1}^{dir}_{e} = false$, there are no i-conflict adjacent links of e in the maximal tree $T^{dir}_{e}(0)$. Let $d(T^{dir}_{e}(0))$ be the length of the longest i-conflict path in $T^{dir}_{e}(0)$.

Theorem 10. *Let $\omega(r) = \sum_{e} |w_e(r) - \phi|$. There exists an algorithm that computes a green time assignment such that $\omega(r + 1) < \omega(r)$ in $O(n \log n)$ time. Further, it attains the optimal value in $O(\sigma n)$ rounds where σ is the standard deviation of the queue lengths in the initial setting.*

Proof. Initially, $g_e(r) = 0$ for each link in E. Let $\mathcal{E} = \{e : 1_e^+ = false \textbf{ or } 1_e^- = false\}$. Consider $e \in \mathcal{E}$ such that w_e is maximum. If $1_e^+ = false$, apply a forward flow f_e^+, and g_e increases and $g_{orth(e)}$ decreases by one. If $1_e^- = false$, apply a forward flow f_e^-, and g_{pred} decreases and $g_{orth(pred(e))}$ increases by one. Consider an incident link e'.

- If $e' \in \{succ(e), borth(e)\}$, $w_{e'}(r) \geq \max(w_{succ(e')}(r), w_{orth(e')}(r))$ and $g_{e'} = 0$, we apply a forward flow, i.e., $f_{e'}^+$.
- If $e' \in \{orth(e), pred(e)\}$, $w_{e'}(r) \geq \max(w_{pred(e')}(r), w_{borth(e')}(r))$ and $g_{pred(e')} = 0$, we apply a forward flow, i.e., $f_{e'}^-$.

Inductively, the process continues until it cannot be extended. Otherwise, another link in M is considered until no more progress can be done. The algorithm is presented in Algorithm 2.

If $\mathcal{E}$ is empty, then the current longest queue length is minimum. Let ϕ be the optimal queue length. Observe that the maximum number of rounds is $\sum_{e:w_e>\phi} w_e - \phi \leq \sum_{e:w_e>\mu} w_e - k$ since $\phi \geq k$. By Chebyshev's inequality, the maximum number of rounds is $an\sigma$ where σ is the standard deviation and a is a constant integer. Therefore, the number of rounds is $O(n\sigma)$.

Algorithm 2: Flooding

1 Let Q be a stack;
2 Let $\mathcal{E} = \{e : 1_e^+ = false \textbf{ or } 1_e^- = false\}$;
3 **while** $M \neq \emptyset$ **do**
4 $\quad$ Let $e \in \mathcal{E}$ such that w_e is maximum;
5 $\quad$ **if** $1_e^+ = false$ **then**
6 $\quad\quad$ $push(Q, (e, +))$;
7 $\quad$ **else if** $1_e^- = false$ **then**
8 $\quad\quad$ $push(Q, (e, -))$;
9 $\quad$ **while** $Q \neq \emptyset$ **do**
10 $\quad\quad$ $(e, dir) \leftarrow pop(Q)$;
11 $\quad\quad$ **if** $dir = `+`$ and $s_e = 0$ and $w_e \geq \max(w_{succ(e)}, w_{orth(e)})$ **then**
12 $\quad\quad\quad$ f_e^+;
13 $\quad\quad\quad$ $push(Q, (succ(e), `+`))$;
14 $\quad\quad\quad$ $push(Q, (orth(e), `-`))$;
15 $\quad\quad$ **else if** $dir = `-`$ and $s_{pred(e)} = 0$ and $w_e \geq \max(w_{pred(e)}, w_{borth(e)})$ **then**
16 $\quad\quad\quad$ f_e^-;
17 $\quad\quad\quad$ $push(Q, (pred(e), `-`))$;
18 $\quad\quad\quad$ $push(Q, (borth(e), `+`))$;

Regarding the complexity, we can compute $\mathcal{E}$ as follows. First, it sorts the links according to the queue length in $O(n \log n)$ time and stores them in an order set. We compute 1_e^+ and 1_e^- starting from the larger until all links have been visited. Therefore, since each link is visited once, we can compute $\mathcal{E}$ in

$O(n \log n)$ time. To compute the green time assignment, we visit each link once. Therefore, the running time algorithm takes $O(n \log n)$. The theorem follows.

6   Local Algorithm

In this section, we show that surprisingly the global problem can be solved locally. In other words, each queue concurrently makes a decision of whether sending a forward or backward flow based only on the queue length of the links at a distance at most two. To break symmetries, we assume that each link has a unique id. The idea is that links at the end of i-conflict path can be reduced safely. However, i-conflict adjacent links can prevent reaching the optimal queue length. We are able to reduce the queue length by keeping the flow direction when links can be part of i-conflict paths.

For the proof we introduce some notation. Let $\max_e^+(r)$ denote the maximum value of the forward neighbors of e and $\max_e^-(r)$ denote the maximum value of the backward neighbors of e. A critical link e is a link with maximum queue length such that either $w_e(r) - 1 = \max_e^+(r)$ or $w_e(r) - 1 = \max_e^-(r)$. Let m be the length of the maximum queue length. Let e, e', e'' be a conflict path. We say that a critical queue is rotating around its conflict cycle if either $w_{e'}(r - 1) = m-1, w_{e''}(r-1) = m, w_e(r) = m-1, w_{e'}(r) = m, w_{e''}(r) = m-1$ and $w_e(r+1) = m, w_{e'}(r) = m - 1$ (see Fig. 5 left) or $w_e(r - 1) = m, w_{e'}(r - 1) = m - 1, w_e(r) = m - 1, w_{e'}(r) = m, w_{e''}(r) = m - 1$ and $w_{e'}(r + 1) = m - 1, w_{e''}(r) = m$ (see Fig. 5 right).

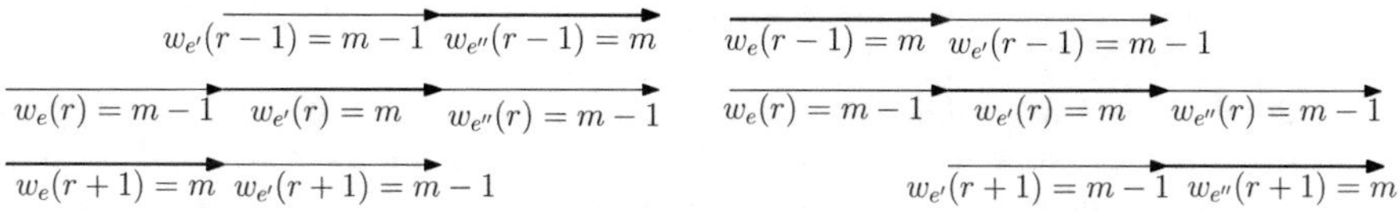

$$w_{e'}(r-1)=m-1 \quad w_{e''}(r-1)=m \qquad w_e(r-1)=m \quad w_{e'}(r-1)=m-1$$
$$w_e(r)=m-1 \quad w_{e'}(r)=m \quad w_{e''}(r)=m-1 \qquad w_e(r)=m-1 \quad w_{e'}(r)=m \quad w_{e''}(r)=m-1$$
$$w_e(r+1)=m \quad w_{e'}(r+1)=m-1 \qquad\qquad w_{e'}(r+1)=m-1 \quad w_{e''}(r+1)=m$$

Fig. 5. Critical queue rotating left and right, respectively.

Let $w_e \succ^2 \max_e^+$ if $w_e > \max_e^+$ and either $w_e > \max(\max_{succ(e)}^+, \max_{orth(e)}^-)$ or, $w_e = \max(\max_{succ(e)}^+, \max_{orth(e)}^-)$ and $id_e > id_{e'}$ for all $e' \in N^+(succ(e)) \cup N^-(orth(e))$ where $w_{e'} = w_e$. Similarly, we define $w_e \succ^2 \max_e^-$.

Theorem 11. *Algorithm 3 minimizes the longest queue length in at most $O(\sigma C_{max}^2)$ rounds where σ is the standard deviation of the queue lengths in the initial setting and C_{max} is the length of the longest conflict cycle with the maximum average queue length.*

Algorithm 3: Local Algorithm (Each link e does in each round)

1 **if** $w_e(r) - 2 > \max_e^+(r)$ **or** $(w_e(r) - 2 = \max_e^+(r)$ **and** $w_e(r) \succ^2 \max_e^+(r))$
   or $(w_e(r) - 1 = \max_e^+(r)$ **and** $w_e(r) \succ^2 \max_e^+(r)$ **and** $s_{(pred(e))}(r-1) \geq 0)$
   then f^+ ;
2 **else if**
   $w_e(r) - 2 > \max_e^-(r)$ **or** $(w_e(r) - 2 = \max_e^-(r)$ **and** $w_e(r) \succ^2 \max_e^-(r))$
   or $(w_e(r) - 1 = \max - +_e(r)$ **and** $w_e(r) \succ^2 \max_e^-(r)$ **and** $s_e(r-1) \leq 0)$
   then f^- ;

---

7 Conclusion

In this paper, we have studied the problem of determining the minimum longest queue length in torus networks. We focus on the fundamental questions of lower bounds and upper bounds on the queue lengths when the agents move in a straight line. We present a global and a local algorithm. However, the number of rounds needed to reach the minimum remains an open problem. Another open problem is to consider asynchronous switches. This paper presents a new perspective on the problem of traffic lights. One direction is to study through simulation the effects when the model becomes probabilistic and, for example, includes turning, rates and agents that appear and disappear.

References

1. Davies, D.W.: The control of congestion in packet-switching networks. IEEE Trans. Commun. **20**(3), 546–550 (1972)
2. Ebrahimi, M., Daneshtalab, M., Liljeberg, P., Plosila, J., Tenhunen, H.: Agent-based on-chip network using efficient selection method. In: 2011 IEEE/IFIP 19th International Conference on VLSI and System-on-Chip (VLSI-SoC), pp. 284–289. IEEE (2011)
3. Ebrahimi, M., Daneshtalab, M., Plosila, J., Tenhunen, H.: MAFA: adaptive fault-tolerant routing algorithm for networks-on-chip. In: 2012 15th Euromicro Conference on Digital System Design (DSD), pp. 201–207. IEEE (2012)
4. Farahnakian, F., Ebrahimi, M., Daneshtalab, M., Liljeberg, P., Plosila, J.: Q-learning based congestion-aware routing algorithm for on-chip network. In: 2011 IEEE 2nd International Conference on Networked Embedded Systems for Enterprise Applications (NESEA), pp. 1–7. IEEE (2011)
5. Jog, A., et al.: OWL: cooperative thread array aware scheduling techniques for improving GPGPU performance. In: ACM SIGPLAN Notices, vol. 48, pp. 395–406. ACM (2013)
6. Kortsarz, G., Peleg, D.: Traffic-light scheduling on the grid. Discret. Appl. Math. **53**(1–3), 211–234 (1994)

Exploiting Modularity of SOS Semantics to Define Quantitative Extensions of Reaction Systems

Linda Brodo[1], Roberto Bruni[2], Moreno Falaschi[3], Roberta Gori[2(✉)],
Francesca Levi[2], and Paolo Milazzo[2]

[1] Dipartimento di Scienze Economiche e Aziendali, Università di Sassari,
Via Muroni 25, Sassari, Italy
brodo@uniss.it
[2] Dipartimento di Informatica, Università di Pisa, Largo B. Pontecorvo 3, Pisa, Italy
{roberto.bruni,roberta.gori,francesca.levi,paolo.milazzo}@unipi.it
[3] Dipartimento di Ingegneria dell'Informazione e Scienze Matematiche,
Università di Siena, Via Roma 56, Siena, Italy
moreno.falaschi@unisi.it

Abstract. Reaction Systems (RSs) are a successful natural computing framework inspired by chemical reaction networks. A RS consists of a set of entities and a set of reactions. Entities can enable or inhibit each reaction, and are produced by reactions or provided by the environment. In a previous paper, we defined an original labelled transition system (LTS) semantics for RSs in the structural operational semantics (SOS) style. This approach has several advantages: (i) it provides a formal specification of the RS dynamics that enables the reuse of many formal analysis techniques and favors the implementation of tools, and (ii) it facilitates the definition of extensions of the RS framework by simply modifying some of the SOS rules in a modular way. In this paper, we demonstrate the extensibility of the framework by defining two quantitative variants of RSs: with reaction delays/durations, and with concentration levels. We provide a prototype logic programming implementation and apply our tool to a RS model of *Th* cells differentiation in the immune system.

Keywords: Bioinformatics · SOS rules · Reaction systems · Logic programming

1 Introduction

Inspired by natural phenomena, many new computational formalisms have been introduced to model different aspects of biology. Basic chemical reactions inspired Reaction Systems (RSs), a *qualitative* modeling formalism introduced

Research supported by University of Pisa PRA_2020_26 *Metodi Informatici Integrati per la Biomedica*, by MIUR PRIN Project 201784YSZ5 *ASPRA–Analysis of Program Analyses*, and by University of Sassari *Fondo di Ateneo per la ricerca 2020*.

C. C. Aranha et al. (Eds.): TPNC 2021, LNCS 13082, pp. 15–32, 2021.
https://doi.org/10.1007/978-3-030-90425-8_2

by Ehrenfeucht and Rozenberg [6,14] that is based on two opposite mechanisms: *facilitation* and *inhibition*. Facilitation means that a reaction can occur only if all its reactants are present, while inhibition means that the reaction cannot occur if any of its inhibitors is present. A *reaction* is hence a triple (R, I, P), where R, I and P are sets of entities representing *reactants*, *inhibitors* and *products*, respectively. A reaction system is represented by a set of reactions having such a form, together with a (finite) support set S containing all of the entities that can appear in reactions. The state of a Reaction System consists of a finite set of entities representing the molecular species that are present in the real system being modeled. Quantities (or concentrations) are not taken into account: the presence of an entity represents the availability of the corresponding molecule in *high concentration*.

The theory of RSs is based on three assumptions: **no permanency**, any entity vanishes unless it is sustained by a reaction; **no counting**, the basic model of RSs is very abstract and qualitative, i.e. the quantity of entities that are present in a cell is not taken into account; **no competition**, an entity is either available for all reactions, or it is not available at all. The computation of the next state of a Reaction System is a deterministic procedure. However, the overall dynamics is influenced by the (set of) contextual entities received (non-deterministically) from the external environment at each step. Such entities join the current state of the system and participate to enabling and disabling reactions. The behaviour of a RS is hence defined as a discrete time interactive process consisting of a *context sequence* (the sets of entities received at each step form the environment), a *result sequence* (the sets of entities produced at each step by reactions) and a *state sequence* (the combined sets of entities present in the system at each step). Since their introduction, RSs have shown to be a quite general computation model whose application ranges from the modeling of biological phenomena [2–4,9], and molecular chemistry [20] to theoretical foundations of computing [12,13].

Labelled Transition Systems (LTSs) are a powerful structure to model the behaviour of interacting processes. In the case of processes described using an algebraic language, LTSs can be conveniently defined following the Structural Operational Semantics (SOS) approach [23]. Given the signature of the language, an SOS system assigns some inference rules to each operator: the conclusion of each rule is the transition of a composite term, which is determined from those of its constituents (appearing as premises of the rule). The SOS approach has been particularly successful in the area of process algebras [15,18,22].

In [8] an LTS semantics in the SOS style for RSs has been proposed. Such a semantics is able to faithfully represent the ordinary semantics of RS, it allows more general experiments to be conducted, and it enriches the expressiveness of contexts. A similar formalization focusing on local scopes for RS entities was proposed in [21]. The SOS approach has several advantages: 1) compositionality: the behaviour of a composite system is defined in term of the behaviours of its constituents; 2) transparency: each transition label conveys information about all the activities connected to the execution step it describes; 3) the notion of contexts is better integrated in the framework; 4) different kinds of contexts (nondeterministic, recursive) are allowed, so to combine different experiments in a single LTS and to account for possibly infinite (regular) computations; 5) extensibility: the definition of enhanced RS variants can be obtained by modifying/adding language operators and SOS rules in a modular fashion; 6) SOS rules facilitate implementation in a declarative language and the use of standard techniques for defining process equivalences.

In this paper we demonstrate the extensibility of the SOS approach to RSs by extending the classical RSs with some relevant quantitative features. First, we add the possibility to express reaction delays and durations. Thanks to this feature we encode reactions with different speeds. Indeed a reaction with associated duration n will deliver its products after n steps. In more details we associate with each reaction a natural number starting from zero (the fastest reaction, the smallest delay, the product being immediately available) up to any value $n > 0$ (slower reactions, greater delay). Following this idea also a duration of 'permanency' can be specified for each reaction. A reaction r that has delay n and duration m will deliver, if applicable, its products after n steps while such products will be available for the following m steps. The second feature which we introduce is some quantitative information that tells how concentrations influence the application of a reaction. This is obtained by adding to each entity in a reaction an approximated quantitative information that will be necessary for enabling the reaction. This feature will allow us to deal with reactions that take into account different levels of concentrations. We note that we still maintain a qualitative perspective on the biological system, since the approximate quantitative information will be used to determine the set of reactions that can be applied in any step whereas competition between different enabled reactions will not be considered. We also provide a freely available prototype implementation in logic programming that allows us to compute and inspect the resulting LTS to perform computational experiments. Finally we apply our tool to the system controlling the differentiation of *Th* cells in the immune system presented in [5, 16].

The structure of the paper is as follows. In Sect. 2 we recall the basics of RSs. In Sect. 3 we recall the syntax and operational semantics of our process algebra for RSs, The original contribution starts from Sect. 4, where we introduce the concepts of delay and duration, and define the corresponding operators. Then, in Sect. 5 we introduce linear functions for expressing the concentration levels of the entities which are necessary to enable a reaction. A prototype implementation in logic programming of our semantic framework is described in Sect. 6 on the basis of a case study about a system controlling the differentiation of *Th* cells in the immune system [5, 16]. Section 7 discusses some related work and concludes the paper.

2   Reaction Systems

The theory of Reaction Systems (RSs) [6] was born in the field of Natural Computing to model the behaviour of biochemical reactions in living cells. While our contribution builds on a process algebraic presentation of RSs, we recall here the main concepts as introduced in the classical set theoretic version. In the following, we use the term *entities* to denote generic molecular substances (e.g., atoms, ions, molecules) that may be present in the states of a biochemical system.

Let S be a (finite) set of entities. A reaction in S is a triple $a = (R, I, P)$, where $R, I, P \subseteq S$ are finite, non empty sets and $R \cap I = \emptyset$. The sets R, I, P are the sets of *reactants*, *inhibitors*, and *products*, respectively. All reactants have to be present in the current state for the reaction to take place. The presence of any of the inhibitors blocks the reaction. Products are the outcome of the reaction, to be released in the next state. We denote with $rac(S)$ the set of all reactions over S. Given $W \subseteq S$, the result of $a = (R, I, P) \in rac(S)$ on W, denoted $res_a(W)$, is given by:

$$res_a(W) \triangleq \begin{cases} P & \text{if } en_a(W) \\ \emptyset & \text{otherwise} \end{cases} \qquad en_a(W) \triangleq R \subseteq W \land I \cap W = \emptyset$$

where $en_a(W)$ is called the *enabling predicate*.

A Reaction System is a pair $\mathcal{A} = (S, A)$ where S is the set of entities, and $A \subseteq rac(S)$ is a finite set of reactions over S. Given $W \subseteq S$, the result of the application of reactions A to W, denoted $res_A(W)$, can be obtained by lifting function res_a to sets of reactions, i.e., $res_A(W) \triangleq \cup_{a \in A} res_a(W)$.

Since living cells are seen as open systems that react to environmental stimuli, the behaviour of a RS is formalized in terms of an *interactive process*. Let $\mathcal{A} = (S, A)$ be a RS and let $n \geq 0$. An n-steps *interactive process* in $\mathcal{A}$ is a pair $\pi = (\gamma, \delta)$ s.t. $\gamma = \{C_i\}_{i \in [0,n]}$ is the *context sequence* and $\delta = \{D_i\}_{i \in [0,n]}$ is the *result sequence*, where $C_i, D_i \subseteq S$ for any $i \in [0, n]$, $D_0 = \emptyset$, and $D_{i+1} = res_A(D_i \cup C_i)$ for any $i \in [0, n-1]$. The context sequence γ represents the environment, while the result sequence δ is entirely determined by γ and A. We call $\tau = W_0, \ldots, W_n$ with $W_i \triangleq C_i \cup D_i$, for any $i \in [0, n]$, the *state sequence*. Note that each state W_i in τ is the union of two sets: the context C_i at step i and the result set $D_i = res_A(W_{i-1})$ from the previous step.

Example 1. We consider a toy RS defined as $\mathcal{A} \triangleq (S, A)$ where $S \triangleq \{a, b, c\}$, and the set of reactions $A \triangleq \{a_1\}$ only contains the reaction $a_1 \triangleq (\{a, b\}, \{c\}, \{b\})$, to be written more concisely as (ab, c, b). Then, we consider a $4-steps$ interactive process $\pi \triangleq (\gamma, \delta)$, where $\gamma \triangleq \{C_0, C_1, C_2, C_3\}$, with $C_0 \triangleq \{a, b\}$, $C_1 \triangleq \{a\}$, $C_2 \triangleq \{c\}$, and $C_3 \triangleq \{c\}$; and $\delta \triangleq \{D_0, D_1, D_2, D_3\}$, with $D_0 \triangleq \emptyset$, $D_1 \triangleq \{b\}$, $D_2 \triangleq \{b\}$, and $D_3 \triangleq \emptyset$. Then, the resulting state sequence is $\tau = W_0, W_1, W_2, W_3 = \{a, b\}, \{a, b\}, \{b, c\}, \{c\}$. In fact, it is easy to check that, e.g., $W_0 = C_0$, $D_1 = res_A(W_0) = res_A(\{a, b\}) = \{b\}$ because $en_a(W_0)$, and $W_1 = C_1 \cup D_1 = \{a\} \cup \{b\} = \{a, b\}$.

3   SOS Rules for Reaction Systems

Inspired by process algebras such as CCS [18], in [8] the authors introduced an algebraic syntax for RSs and equipped it with SOS inference rules defining the behaviour of each operator. This allows us to consider a LTS semantics for RSs, where states are terms of the algebra, each transition corresponds to a step of the RS and transition labels retain some information on the entities needed to perform each step.

Definition 2 (RS processes). *Let S be a set of entities. An* RS process P *is any term defined by the following grammar:*

$$\mathsf{P} := [\mathsf{M}] \qquad \mathsf{M} := (R, I, P) \mid D \mid \mathsf{K} \mid \mathsf{M}|\mathsf{M} \qquad \mathsf{K} ::= \mathbf{0} \mid X \mid C.\mathsf{K} \mid \mathsf{K} + \mathsf{K} \mid \mathsf{rec}\ X.\ \mathsf{K}$$

where $R, I, P \subseteq S$ are non empty sets of entities, $C, D \subseteq S$ are possibly empty set of entitities, and X is a process variable.

An RS process P embeds a *mixture* process M obtained as the parallel composition of some reactions (R, I, P), some set of currently present entities D (possibly the empty set $\emptyset$), and some *context* process K. We write $\prod_{i \in I} \mathsf{M}_i$ for the parallel composition of all M_i with $i \in I$. For example, $\prod_{i \in \{1,2\}} \mathsf{M}_i = \mathsf{M}_1 \mid \mathsf{M}_2$.

A process context K is a possibly nondeterministic and recursive system: the nil context $\mathbf{0}$ stops the computation; the prefixed context $C.\mathsf{K}$ says that the entities in C are immediately available to be consumed by the reactions, and then K is the context offered at the next step; the non deterministic choice $\mathsf{K}_1 + \mathsf{K}_2$ allows the context to behave either as K_1 or K_2; X is a process variable, and rec $X.$ K is the usual recursive operator of process algebras. We write $\sum_{i \in I} \mathsf{K}_i$ for the nondeterministic choice between all K_i with $i \in I$.

We say that P and P' are structurally equivalent, written $\mathsf{P} \equiv \mathsf{P}'$, when they denote the same term up to the laws of commutative monoids (unit, associativity and commutativity) for parallel composition $\cdot | \cdot$, with $\emptyset$ as the unit, and the laws of idempotent and commutative monoids for choice $\cdot + \cdot$, with $\mathbf{0}$ as the unit. We also assume $D_1 | D_2 \equiv D_1 \cup D_2$ for any $D_1, D_2 \subseteq S$.

Remark 3. Note that the processes $\emptyset$ and $\mathbf{0}$ are not interchangeable: as it will become clear from the operational semantics, the process $\emptyset$ can perform just a trivial transition to itself, while the process $\mathbf{0}$ cannot perform any transition and can be used to stop the computation.

Definition 4 (RSs as RS processes). *Let $\mathcal{A} = (S, A)$ be a RS, and $\pi = (\gamma, \delta)$ an n-step interactive process in $\mathcal{A}$, with $\gamma = \{C_i\}_{i \in [0,n]}$ and $\delta = \{D_i\}_{i \in [0,n]}$. For any step $i \in [0, n]$, the corresponding RS process $[\![\mathcal{A}, \pi]\!]_i$ is defined as follows:*

$$[\![\mathcal{A}, \pi]\!]_i \triangleq \left[\prod_{a \in A} a \mid D_i \mid \mathsf{K}_{\gamma^i} \right]$$

where the context process $\mathsf{K}_{\gamma^i} \triangleq C_i.C_{i+1}.\cdots.C_n.\mathbf{0}$ is the sequentialization of the entities offered by γ^i (the shifting of γ starting at the i-th step). We write $[\![\mathcal{A}, \pi]\!]$ as a shorthand for $[\![\mathcal{A}, \pi]\!]_0$.

Example 5. Here, we give the encoding of the reaction system, $\mathcal{A} = (S, A)$, defined in Example 1. The resulting RS process is as follows:

$$\mathsf{P} \triangleq [\![\mathcal{A}, \pi]\!] = [\![(\{\mathsf{a}, \mathsf{b}, \mathsf{c}\}, \{(\mathsf{ab}, \mathsf{c}, \mathsf{b})\}), \pi]\!] = [(\mathsf{ab}, \mathsf{c}, \mathsf{b}) \mid \emptyset \mid \mathsf{K}_\gamma] \equiv [(\mathsf{ab}, \mathsf{c}, \mathsf{b}) \mid \mathsf{K}_\gamma]$$

where $\mathsf{K}_\gamma = \{\mathsf{a}, \mathsf{b}\}.\{\mathsf{a}\}.\{\mathsf{c}\}.\{\mathsf{c}\}.\mathbf{0}$, written more concisely as $\mathsf{ab.a.c.c.0}$. Note that $D_0 = \emptyset$ is inessential and can be discarded thanks to structural congruence.

In Definition 4 we have not exploited the entire potentialities of the syntax. In particular, the context K_γ is just a finite sequence of action prefixes induced by the set of entities provided by γ at the various steps. Our syntax allows for more general kinds of contexts as shown in the example below. Nondeterminstic contexts can be used to descrive several alternative experimental conditions, while recursion can be exploited to extract some regularity in the longterm behaviour of a RS. Together, they can deal with any combination of in-breadth/in-depth behavioural analysis.

Example 6. Let us further elaborate on our running example. Suppose we want to enhance the behaviour of the context by defining a process $\mathsf{K}' \triangleq \mathsf{K}_1 + \mathsf{K}_2$ that non-deterministically can behave as either $\mathsf{K}_1 \triangleq \mathsf{ab.a.c.c.0}$ (as in Example 5), or $\mathsf{K}_2 \triangleq \mathrm{rec}\ X.\ \mathsf{ab.a}.X$ (which is a recursive behaviour that allows the reaction to be always enabled). Then we simply set $\mathsf{P}' \equiv [(\mathsf{ab}, \mathsf{c}, \mathsf{b}) \mid \mathsf{K}']$.

Definition 7 (Label). *A label is a tuple $\langle W \rhd R, I, P \rangle$ with $W, R, I, P \subseteq S$. The set of transition labels is ranged over by ℓ.*

In a transition label $\langle W \rhd R, I, P \rangle$, we record the set W of entities currently in the system (produced in the previous step or provided by the context), the set R of entities whose presence is assumed (either because they are needed as reactants on an applied reaction or because their presence prevents the application of some reaction); the set I of entities whose absence is assumed (either because they appear as inhibitors for an applied reaction or because their absence prevents the application of some reaction); the set P of products of all the applied reactions.

Definition 8 (Operational semantics). *The operational semantics of processes is defined by the set of SOS inference rules in Fig. 1.*

The process $\mathbf{0}$ has no transition. The rule *(Ent)* makes available the entities in the (possibly empty) set D, then reduces to $\emptyset$. As a special instance of *(Ent)*, $\emptyset \xrightarrow{\langle \emptyset \rhd \emptyset, \emptyset, \emptyset \rangle} \emptyset$. The rule *(Cxt)* says that a prefixed context process $C.\mathsf{K}$ makes available the entities in the set C and then reduces to K. The rule *(Rec)* is the classical rule for recursion. Here, $\mathsf{K}[^{\mathrm{rec}\ X.\ \mathsf{K}}/_X]$ denotes the process obtained by replacing in K every free occurrence of the variable X with its recursive definition $\mathrm{rec}\ X.\ \mathsf{K}$. For example $\mathrm{rec}\ X.\ \mathsf{a.b}.X \xrightarrow{\langle \mathsf{a} \rhd \emptyset, \emptyset, \emptyset \rangle} \mathsf{b.rec}\ X.\ \mathsf{a.b}.X$ The rules *(Suml)* and *(Sumr)* select a move of either the left or the right component, resp., discarding the other process. The rule *(Pro)*, executes the reaction (R, I, P) (its reactants,

$$\frac{}{D \xrightarrow{\langle D \triangleright \emptyset, \emptyset, \emptyset \rangle} \emptyset} \ (Ent) \qquad \frac{}{C.K \xrightarrow{\langle C \triangleright \emptyset, \emptyset, \emptyset \rangle} K} \ (Cxt) \qquad \frac{K_1 \xrightarrow{\ell} K_1'}{K_1 + K_2 \xrightarrow{\ell} K_1'} \ (Suml) \qquad \frac{K_2 \xrightarrow{\ell} K_2'}{K_1 + K_2 \xrightarrow{\ell} K_2'} \ (Sumr)$$

$$\frac{}{(R, I, P) \xrightarrow{\langle \emptyset \triangleright R, I, P \rangle} (R, I, P) \mid P} \ (Pro) \qquad \frac{J \subseteq I \quad Q \subseteq R \quad J \cup Q \neq \emptyset}{(R, I, P) \xrightarrow{\langle \emptyset \triangleright J, Q, \emptyset \rangle} (R, I, P)} \ (Inh) \qquad \frac{K[^{\mathrm{rec}\ X.K}/x] \xrightarrow{\ell} K'}{\mathrm{rec}\ X.\ K \xrightarrow{\ell} K'} \ (Rec)$$

$$\frac{M_1 \xrightarrow{\ell_1} M_1' \quad M_2 \xrightarrow{\ell_2} M_2' \quad \ell_1 \frown \ell_2}{M_1 \mid M_2 \xrightarrow{\ell_1 \cup \ell_2} M_1' \mid M_2'} \ (Par) \qquad \frac{M \xrightarrow{\langle W \triangleright R, I, P \rangle} M' \quad R \subseteq W}{[M] \xrightarrow{\langle W \triangleright R, I, P \rangle} [M']} \ (Sys)$$

Fig. 1. SOS semantics of the RS processes.

inhibitors, and products are recorded the label), which remains available at the next step together with P. The rule (Inh) applies when the reaction (R, I, P) should not be executed; it records in the label the possible causes for which the reaction is disabled: possibly some inhibiting entities ($J \subseteq I$) are present or some reactants ($Q \subseteq R$) are missing, with $J \cup Q \neq \emptyset$, as at least one cause is needed for explaining why the reaction is not enabled. The rule (Par) puts two processes in parallel by pooling their labels and joining all the set components of the labels. The sanity check $\ell_1 \frown \ell_2$ is required to guarantee that there is no conflict between reactants and inhibitors of the applied reactions:

$$\langle W_1 \triangleright R_1, I_1, P_1 \rangle \frown \langle W_2 \triangleright R_2, I_2, P_2 \rangle \triangleq (W_1 \cup W_2 \cup R_1 \cup R_2) \cap (I_1 \cup I_2) = \emptyset$$

In the conclusion of rule (Par) we write $\ell_1 \cup \ell_2$ for the componentwise union of labels:

$$\langle W_1 \triangleright R_1, I_1, P_1 \rangle \cup \langle W_2 \triangleright R_2, I_2, P_2 \rangle \triangleq \langle W_1 \cup W_2 \triangleright R_1 \cup R_2, I_1 \cup I_2, P_1 \cup P_2 \rangle$$

Finally, the rule (Sys) requires that all the processes of the systems have been considered, and also checks that all the needed reactants are actually available in the system ($R \subseteq W$). In fact this constraint can only be met on top of all processes. The check that inhibitors are absent ($I \cap W = \emptyset$) is not necessary, as it is embedded in rule (Par).

Example 9. Let us consider the RS process $\mathsf{P}_0 \triangleq [(\mathsf{ab}, \mathsf{c}, \mathsf{b}) \mid \mathsf{ab}.\mathsf{a}.\mathsf{c}.\mathsf{c}.\mathsf{0}]$ from Example 5. The process P_0 has a unique outgoing transition, whose formal derivation is given below:

$$\frac{\dfrac{}{(\mathsf{ab}, \mathsf{c}, \mathsf{b}) \xrightarrow{\langle \emptyset \triangleright \mathsf{ab}, \mathsf{c}, \mathsf{b} \rangle} (\mathsf{ab}, \mathsf{c}, \mathsf{b}) \mid \mathsf{b}} (Pro) \quad \dfrac{}{\mathsf{ab}.\mathsf{a}.\mathsf{c}.\mathsf{c}.\mathsf{0} \xrightarrow{\langle \mathsf{ab} \triangleright \emptyset, \emptyset, \emptyset \rangle} \mathsf{a}.\mathsf{c}.\mathsf{c}.\mathsf{0}} (Cxt)}{\dfrac{(\mathsf{ab}, \mathsf{c}, \mathsf{b}) \mid \mathsf{ab}.\mathsf{a}.\mathsf{c}.\mathsf{c}.\mathsf{0} \xrightarrow{\langle \mathsf{ab} \triangleright \mathsf{ab}, \mathsf{c}, \mathsf{b} \rangle} (\mathsf{ab}, \mathsf{c}, \mathsf{b}) \mid \mathsf{b} \mid \mathsf{a}.\mathsf{c}.\mathsf{c}.\mathsf{0}}{[(\mathsf{ab}, \mathsf{c}, \mathsf{b}) \mid \mathsf{ab}.\mathsf{a}.\mathsf{c}.\mathsf{c}.\mathsf{0}] \xrightarrow{\langle \mathsf{ab} \triangleright \mathsf{ab}, \mathsf{c}, \mathsf{b} \rangle} [(\mathsf{ab}, \mathsf{c}, \mathsf{b}) \mid \mathsf{b} \mid \mathsf{a}.\mathsf{c}.\mathsf{c}.\mathsf{0}]} (Sys)} (Par)$$

The target process $\mathsf{P}_1 \triangleq [(\mathsf{ab}, \mathsf{c}, \mathsf{b}) \mid \mathsf{b} \mid \mathsf{a}.\mathsf{c}.\mathsf{c}.\mathsf{0}]$ has also a unique outgoing transition, namely:

$$\mathsf{P}_1 = [(\mathsf{ab}, \mathsf{c}, \mathsf{b}) \mid \mathsf{b} \mid \mathsf{a}.\mathsf{c}.\mathsf{c}.\mathsf{0}] \xrightarrow{\langle \mathsf{ab} \triangleright \mathsf{ab}, \mathsf{c}, \mathsf{b} \rangle} [(\mathsf{ab}, \mathsf{c}, \mathsf{b}) \mid \mathsf{b} \mid \mathsf{c}.\mathsf{c}.\mathsf{0}] = \mathsf{P}_2$$

Instead the process P_2 has three outgoing transitions, each providing a different justification to the fact that the reaction (ab, c, b) is not enabled. Notably, the three transitions have the same target process $P_3 \triangleq [(ab, c, b) \mid c.\mathbf{0}]$.

1. $P_2 \xrightarrow{\langle bc \triangleright c, \emptyset, \emptyset \rangle} P_3$ shows that the presence of c has played some role in inhibiting the reaction;

2. $P_2 \xrightarrow{\langle bc \triangleright \emptyset, a, \emptyset \rangle} P_3$ shows that the absence of a has played some role in inhibiting the reaction.

3. $P_2 \xrightarrow{\langle bc \triangleright c, a, \emptyset \rangle} P_3$ shows that the presence of c and the absence of a inhibited the reaction; this label is thus more informative than the previous two (in a sense formalised in [8]).

Finally, the process P_3 has seven transitions all leading to $P_4 \triangleq [(ab, c, b) \mid \mathbf{0}]$. Their labels are of the form $\langle c \triangleright J, Q, \emptyset \rangle$ with $J \subseteq c$, $Q \subseteq ab$ and $J \cup Q \neq \emptyset$. Each label provides a different explanation why the reaction is not enabled.

The following theorem from [8] shows that the rewrite steps of a RS exactly match the transitions of its corresponding RS process.

Theorem 10. *Let $\mathcal{A} = (S, A)$ be a RS, and $\pi = (\gamma, \delta)$ an n-step interactive process in $\mathcal{A}$ with $\gamma = \{C_i\}_{i \in [0,n]}$, $\delta = \{D_i\}_{i \in [0,n]}$, and let $W_i \triangleq C_i \cup D_i$ and $P_i \triangleq [\![\mathcal{A}, \pi]\!]_i$ for any $i \in [0, n]$. Then:*

1. $\forall i \in [0, n-1]$, $P_i \xrightarrow{\langle W \triangleright R, I, P \rangle} P$ implies $W = W_i$, $P = D_{i+1}$ and $P \equiv P_{i+1}$;

2. $\forall i \in [0, n-1]$, there exists $R, I \subseteq S$ such that $P_i \xrightarrow{\langle W_i \triangleright R, I, D_{i+1} \rangle} P_{i+1}$.

Remark 11. Note that the process $P_n = [\![\mathcal{A}, \pi]\!]_n = [\prod_{a \in A} a \mid D_n \mid C_n.\mathbf{0}]$ has one more transition available (the $(n+1)$-th step from P_0), even if the standard theory of RSs stops the computation after n steps. We thus have additional steps

$$P_n \xrightarrow{\langle W_n \triangleright R_n, I_n, res_A(W_n) \rangle} \left[\prod_{a \in A} a \mid res_A(W_n) \mid \mathbf{0} \right]$$

for suitable $R_n, I_n \subseteq S$. The target process contains $\mathbf{0}$ and therefore is deadlock.

Example 9 shows that we can have redundant transitions because of rule *(Inh)*. However, they can be easily detected and eliminated by considering a notion of dominance [8].

4   Delays and Durations

In Biology it is well known that reactions occur with different frequencies. For example, since enzymes catalyze reactions, many reactions are more frequent when some enzymes are present, and less frequent when such enzymes are absent. Moreover, reactions describing complex transformations may require time before releasing their products. To capture these dynamical aspects in our framework by preserving the discrete and abstract nature of RS, we propose a discretization of the *delay* between two occurrences of a reaction by using a scale of natural numbers, from 0 (smallest delay, highest frequency) up to n (increasing delay, lower frequency).

Intuitively, the notation D^n stands for making the entities D available after n time units, and we use the shorthand D for D^0. Similarly, we can associate a delay value to the product of each reaction by writing $(R, I, P)^n$ when the product of the reaction will be available after n time units, and we write (R, I, P) for $(R, I, P)^0$. The syntax for mixture processes is thus extended as below and the operational semantics is changed accordingly (see Fig. 2).

$$M ::= (R, I, P)^n \mid D^n \mid K \mid M|M$$

Rule (*Tick*) represents the passing of one time unit, while rule (*Set*) notify the availability of entities whose delay has expired. Rule (*ProS*) attaches to the product of the reaction the same delay as the one of the reaction itself, while rule (*InhS*) is used when the reaction is not enabled.

$$\frac{}{D \xrightarrow{\langle D \triangleright \emptyset, \emptyset, \emptyset \rangle} \emptyset} \; (Set) \qquad \frac{}{D^{n+1} \xrightarrow{\langle \emptyset \triangleright \emptyset, \emptyset, \emptyset \rangle} D^n} \; (Tick)$$

$$\frac{}{(R, I, P)^n \xrightarrow{\langle \emptyset \triangleright R, I, P \rangle} (R, I, P)^n \mid P^n} \; (ProS) \qquad \frac{J \subseteq I \quad Q \subseteq R \quad J \cup Q \neq \emptyset}{(R, I, P)^n \xrightarrow{\langle \emptyset \triangleright J, Q, \emptyset \rangle} (R, I, P)^n} \; (InhS)$$

Fig. 2. SOS semantics with delays and durations.

Note that the context definition is unchanged. The encoding described in Definition 4 still applies.

Example 12. Let us consider two RSs sharing the same entity set $S = \{a, b, c, d\}$ and the same reactions $a_1 = (a, b, b)$, $a_2 = (b, a, a)$, $a_3 = (ac, b, d)$, $a_4 = (d, a, c)$, but working with different reaction speeds. For simplicity we assume only two speed levels are distinguished: 0 the fastest and 1 the slowest. The reaction system P_1 provides the following speed assignment to the reactions: $\{a_1^1, a_2, a_3, a_4^1\}$. The reaction system P_2 provides the following speed assignment to the reactions: $\{a_1, a_2^1, a_3^1, a_4\}$. We assume that the context process for both reaction systems is just $ac.\emptyset.0$:

$$P_1 = [ac.\emptyset.0|a_1^1|a_2|a_3|a_4^1] \xrightarrow{\langle ac \triangleright ac, bd, bd \rangle} [\emptyset.0|b^1|d|a_1^1|a_2|a_3|a_4^1] \xrightarrow{\langle d \triangleright d, abc, c \rangle} [0|c^1|b|a_1^1|a_2|a_3|a_4^1]$$

$$P_2 = [ac.\emptyset.0|a_1|a_2^1|a_3^1|a_4] \xrightarrow{\langle ac \triangleright ac, bd, bd \rangle} [\emptyset.0|b|d^1|a_1|a_2^1|a_3^1|a_4] \xrightarrow{\langle b \triangleright b, acd, a \rangle} [0|a^1|d|a_1|a_2^1|a_3^1|a_4]$$

Additionally, inspired by [7], we can also provide entities with a duration, i.e. entities that last a finite number of steps. To this aim we use the syntax $D^{[n,m]}$

to represent the availability of D for m time units starting after n time units from the current time. By assuming that each reaction only produces entities with the same duration, we can describe duration and delay also associated to reactions: $(R, I, P)^{[n,m]}$ means that all the entities in P (the products) have a delay of n but will last m steps (once they appear in the state). While we could easily define the SOS rules for the above processes, we note that durations are just syntax sugar: we can simply let

$$D^{[n,m]} \triangleq \prod_{k=n}^{n+m} D^k \qquad (R, I, P)^{[n,m]} \triangleq \prod_{k=n}^{n+m} (R, I, P)^k.$$

For example, we have $\mathsf{a}^{[2,2]} = \mathsf{a}^2|\mathsf{a}^3|\mathsf{a}^4$ and $\mathsf{a}^{[0,0]} = \mathsf{a}^0 = \mathsf{a}$.

Example 13. The cell cycle is a series of sequential events leading to cell duplication. It consists of four phases: G_1, S, G_2 and M. The first three phases (G_1, S, and G_2) are called interphase (I). In these phases, the main event which happens is the replication of DNA. In the last phase (M), called mitosis, the cell segregates the duplicated sets of chromosomes between daughter cells, and then divides. The duration of the cell cycle depends on the type of cell.

In [24] a Delay Differential Equation model of tumour growth has been proposed, that includes the immune system response and a phase-specific drug able to alter the natural course of action of the cell cycle of the tumour cells. A delay is used to model the duration of the interphase.

Inspired from [24] we define a RS model of tumour growth using delays and durations. We consider two populations of tumour cells: those in the interphase of the cell cycle ($\mathsf{T_I}$) and those in mitosis phase ($\mathsf{T_M}$). We assume that cells reside in the interphase for σ time units. Moreover, we represent the drug with entity D and assume that, once received from the environment, it takes an active form $\mathsf{D_a}$ and disappears after a delay of δ time units. The reactions of the model are the following: $\mathsf{a_1} = (\mathsf{T_I}, \mathsf{D_a}, \mathsf{T_M})^\sigma$, $\mathsf{a_2} = (\mathsf{T_M}, \emptyset, \mathsf{T_I})$, $\mathsf{a_3} = (\mathsf{D}, \emptyset, \mathsf{D_a})^{[0,\delta]}$. Let $A = \mathsf{a_1}|\mathsf{a_2}|\mathsf{a_3}$.

Let us assume that the system starts from a configuration in which tumour cells are in the interphase. Hence, the RS process is $P = [\mathsf{K}|\mathsf{T_I}|A]$, where K is a contex process. Now, by providing different definitions for K we can emulate different drug administration strategies. For instance, let us consider $\sigma = 1$, $\delta = 1$ and these two context processes:

– $\mathsf{K_1} = \mathsf{rec}\ X. \emptyset.X$ (i.e., drug not administered)
– $\mathsf{K_2} = \mathsf{rec}\ X. \mathsf{D}.\emptyset.\emptyset.\emptyset.X$ (i.e., drug administered every 4 time units)

Now, when no drug is administered, tumour cells execute the cell cycle infinitely:

$$P[^{K_1}/_K] = [\mathsf{K_1}|\mathsf{T_I}|A] \xrightarrow{\langle \mathsf{T_I} \rhd \mathsf{T_I}, \mathsf{D_a} \mathsf{T_M} \mathsf{D}, \mathsf{T_M}\rangle} [\mathsf{K_1}|\mathsf{T_M}^1|A] \xrightarrow{\langle \emptyset \rhd \emptyset, \mathsf{T_I} \mathsf{T_M} \mathsf{D}, \emptyset \rangle}$$
$$[\mathsf{K_1}|\mathsf{T_M}|A] \xrightarrow{\langle \mathsf{T_M} \rhd \mathsf{T_M}, \mathsf{T_I} \mathsf{D}, \mathsf{T_I}\rangle} [\mathsf{K_1}|\mathsf{T_I}|A] \xrightarrow{\langle \mathsf{T_I} \rhd \mathsf{T_I}, \mathsf{D_a} \mathsf{T_M} \mathsf{D}, \mathsf{T_M}\rangle} \ldots$$

When the drug is administered every 4 time units, the cell cycle is interrupted after a few steps:

$$P[{}^{K_2}/_K] = [K_2|T_I|A] \xrightarrow{\langle T_I D \rhd T_I D, D_a T_M, T_M D_a \rangle} [\emptyset.\emptyset.\emptyset.K_2|D_a^1|D_a|T_M^1|A] \xrightarrow{\langle D_a \rhd D_a, T_I T_M D, \emptyset \rangle}$$

$$[\emptyset.\emptyset.K_2|D_a|T_M|A] \xrightarrow{\langle D_a T_M \rhd D_a T_M, T_I D, T_I \rangle} [\emptyset.K_2|T_I|A] \xrightarrow{\langle T_I \rhd T_I, D_a T_M D, T_M \rangle}$$

$$[K_2|T_M^1|A] \xrightarrow{\langle D \rhd D, T_I T_M, D_a \rangle} [\emptyset.\emptyset.\emptyset.K_2|D_a^1|D_a|T_M|A] \xrightarrow{\langle T_M D_a \rhd D_a T_M, T_I D, T_I \rangle} [\emptyset.\emptyset.K_2|D_a|T_I|A] \,.$$

Alternative drug scheduling could be tested by providing alternative definitions for K.

5   Concentration Levels Through Linear Functions

Quantitative modelling of chemical reaction requires taking molecule concentrations into account. An abstract representation of concentrations that is considered in many formalisms is based on *concentration levels*: rather than representing such quantities as real numbers, a finite classification is considered (e.g., *low/medium/high*) with a granularity that reflects the number of concentrations levels at which significant changes in the behaviour of the molecule are observed. In classical RSs, the modelling of concentration levels would require using different entities for the same molecule (e.g., a_1, a_m, and a_h for low, medium and high concentration of a, respectively). This may introduce some additional complexity due to the need of guaranteeing that only one of these entities is present at any time for the state to be consistent. Moreover, consistency would be put at risk also by the fact that entities representing different levels of the same molecule (e.g., a_1 and a_h) could be provided at the same time by the context.

We now enhance RS process by adding some quantitative information associated to each entity of each reaction, so that levels are just natural numbers and the concentration levels of the products depend on the concentration levels of reactants. The idea is to associate linear expressions (such as $e = m \cdot x + n$, with $m \in \mathbb{N}$ and $n \in \mathbb{N}^+$)[1] to reactants and products of each reaction (we write $s(e)$ to state that expression e is associated to entity s). Expressions associated to reactants are used as *patterns* to match the current levels of the entities involved in the reaction. Pattern matching allows variable x (the same for all reactants) to be instantiated with a value that depends on the levels. Then, linear expressions associated to products (that can contain, again, variable x) can be evaluated to compute the concentration levels of those entities. Expressions can be associated also to reaction inhibitors in order let such entities inhibit the reaction only when their concentration level is above a given threshold. For expressions associated to inhibitors, we will require them to be *closed*, namely they cannot contain the $m \cdot x$ term and simply correspond to a positive natural number.

Also the state of the system has to take into account concentration levels. Consequently, in the definition of states we will exploit again closed expressions

---

[1] To ease the presentation, we impose $n \in \mathbb{N}^+$ on the linear epression e to guarantee that its evaluation into a positive number, even when $x = 0$. Alternative choices are possible to relax this constraint.

to obtain that each entity in the a state is associated to a natural number representing its concentration level.

Example 14. Assume that we want to write a reaction that produces c with a concentration level that corresponds to the current concentration level of a, and that requires b not to be present at a concentration level higher than 1. Such a reaction would be $r = (R, I, P)$ where $R = a(x + 1)$, $I = b(2)$ and $P = c(x + 1)$. In the state $\{a(3), b(1)\}$. Reaction r is enabled by taking $x = 2$ (the maximum value for x that satisifes $x + 1 \leq 3$). Since $b(1) < b(2)$, entity c will be produced with concentration level $(x + 1) = 3$. On the contrary, in the state $\{a(2), b(2)\}$ the reaction a is not enabled because the concentration of the inhibitor is too high.

To formalize the above linear constraints we introduce some notations. A *closed* linear expression is just a natural number, and $e[v/x]$ represents the *substitution* of variable x with the value v in e. A *pattern* $p = \{s_1(e_1), ..., s_k(e_k)\}$ is a set of associations of linear expressions to entities. We write $p(s_i)$ for the linear expression associated with s_i in p. A pattern p is closed if $p(s_i) \in \mathbb{N}$ for any $s_i \in S$ and we write $p[v/x]$ to mean the closed pattern obtained as $p[v/x](s_i) = e_i[v/x]$ for all s_i. Given two closed patterns p, q, we write $p \leq q$ if $p(s) \leq q(s)$ for all $s \in S$.

Then, we extend the syntax of reactions $r = (R, I, P)$ by considering I as closed pattern, and R and P as patterns such that if P is not closed, then R is not closed. A state W is a closed pattern. At each step, starting from a given state, the semantics verifies the enabled reactions using function $en()$, computes the *multiplicity* of reaction applications (the value of x obtained by matching the current state W with pattern R) by function $mul()$, and computes the resulting state by function $res()$. Given a reaction $a = (R, I, P)$ and a state W, we define:

– the function $en(a, W)$, returns 1 if the reaction is enabled, 0 otherwise

$$en(a, W) \triangleq \begin{cases} 1 & \text{if } R[0/x] \leq W \text{ and } \forall s \in S. \ I(s) > 0 \Rightarrow W(s) < I(s) \\ 0 & \text{otherwise} \end{cases}$$

– function $mul(a, W)$ returns the value v that will correctly bind x when applied to state W

$$mul(a, W) \triangleq \begin{cases} 1 & \text{if } en(a, W) = 0 \text{ or } R \text{ is a closed pattern} \\ \max\{v \in \mathbb{N} \mid R[v/x] \leq W\} & \text{otherwise} \end{cases}$$

– function $res(a, W)$ returns the product of the application of reaction a on state W

$$res(a, W) \triangleq en(a, W) \cdot P[mul(a, W)/x]$$

Example 15. Consider again the previous example, $r = (R, I, P)$ with $R = a(x + 1)$, $I = b(2)$ and $P = c(x + 1)$ and the state $W = \{a(3), b(1)\}$, we compute:

– $en(r, W) = 1$, as $R[0/x] = a(1) \leq a(3)$ and $W(b) = 1 < 2 = I(b)$.
– $mul(r, W) = 2$, as $\max\{x \in \mathbb{N} \mid R[v/x] = a(v + 1) \leq a(3)b(1) = W\} = 2$.

$-$ $res(r, W) = en(r, W) \cdot P[mul(r, W)/x] = 1 \cdot \mathsf{c}(2 + 1) = \mathsf{c}(3)$.

Once the product of each enabled reaction has been calculated, we need to compute the next state. We consider the operator that computes the maximum between two closed patterns $p \sqcup q$, defined as $(p \sqcup q)(s) = \max\{p(s), q(s)\}$. It gives the point-wise maximum value of each entity. Analogously, to combine inhibitor constraints, we will later consider the operator that computes the minimum between two closed patterns p and q, denoted by $p \sqcap q$, defined as $(p \sqcap q)(s) = \min\{p(s), q(s)\}$.

Example 16. Assume we add a new reaction $r' = (R', I', P')$ to the previous example, where $R' = \mathsf{a}(x + 2)\mathsf{b}(1)$, $I' = \emptyset$, $P' = \mathsf{c}(3x + 2)$. By applying the function $res(r', W) = en(r', W) \cdot P[mul(r', W)/x]$ we have $1 \cdot \mathsf{c}(3x + 2)[1/x] = \mathsf{c}(5)$. Therefore, in the system composed by state W and reactions r, and r', we obtain the next state $W' = \mathsf{c}(3) \sqcup \mathsf{c}(5) = \mathsf{c}(5)$.

In the SOS style, the hypotheses under which a reaction is applied or inhibited are recorded in the label and their consistency is verified by rule (Par) and (Sys). We stretch here the fact that such hypotheses consist of constraints over concentration levels. If we assume that a reaction $a = (R, I, P)$ is enabled with multiplicity v, it means that it must be $\forall s \in I$. $W(s) < I(s)$ and $\forall s \in S$. $R[v/x](s) \le W(s)$ but $R[v + 1/x] \not\le W$. The first two constraints can be already represented in the ordinary labels, for the last one we extend labels with a set of bounds $c = \{R_1, ..., R_n\}$ for which we shall require that $\forall i \in [1, n]$. $R_i \not\le W$, (more concisely $c \not\le W$) and let

$$bnd(R, v) \triangleq \begin{cases} \emptyset & \text{if } R \text{ is a closed pattern} \\ \{R[v + 1/x]\} & \text{otherwise} \end{cases}$$

Correspondingly, we update the operation to combine and to compare labels as follows:

$$\langle W_1 \rhd R_1, I_1, P_1 \rangle \otimes \langle W_2 \rhd R_2, I_2, P_2 \rangle \triangleq \langle W_1 \sqcup W_2 \rhd R_1 \sqcup R_2, I_1 \sqcap I_2, P_1 \sqcup P_2 \rangle$$

$$\langle W_1 \rhd R_1, I_1, P_1 \rangle \frown \langle W_2 \rhd R_2, I_2, P_2 \rangle \triangleq \forall s \in S. \ I(s) > 0 \Rightarrow WR(s) < I(s)$$

where $I = I_1 \sqcap I_2$ and $WR = W_1 \sqcup W_2 \sqcup R_1 \sqcup R_2$.

$$\frac{v \in \mathbb{N} \quad R_v = R[v/x] \quad P_v = P[v/x] \quad c = bnd(R, v)}{(R, I, P) \xrightarrow{c, \langle \emptyset \rhd R_v, I, P_v \rangle} (R, I, P) \mid P_v} (ProS)$$

$$\frac{J \le I \quad Q \le R[0/x] \quad J \sqcup Q \ne \emptyset}{(R, I, P) \xrightarrow{\emptyset, \langle \emptyset \rhd J, Q, \emptyset \rangle} (R, I, P)} (InhS)$$

$$\frac{M_1 \xrightarrow{c_1, \ell_1} M_1' \quad M_2 \xrightarrow{c_2, \ell_2} M_2' \quad \ell_1 \frown \ell_2}{M_1 \mid M_2 \xrightarrow{c_1 \cup c_2, \ell_1 \otimes \ell_2} M_1' \mid M_2'} (Par)$$

$$\frac{M \xrightarrow{c, \langle W \rhd R, I, P \rangle} M' \quad c \not\le W \quad R \le W}{[M] \xrightarrow{c, \langle W \rhd R, I, P \rangle} [M']} (Sys)$$

Fig. 3. SOS semantics for linear functions.

6   Application

In [8] we presented a preliminary implementation of RSs in a logic programming language (Prolog), mainly intended for rapid prototyping. Following the formal definition of RS with delays/durations and with concentration levels given in the previous sections, we describe how such extensions have been integrated in the prototype, available for download[2]. We remark that the modular nature of the SOS formalization simplified significantly the adaptation of the tool.

A RS is represented as a list of reactions. In the case of delays/durations, a reaction is a quintuple [R, I, P, N, M], while in the case of concentration levels it is a triple [R, I, P]. There, R, I and P represent reactants, inhibitors and products, respectively, N is the delay and M the duration. After coding the RS, a query is performed by calling either the predicate `computation(InitialState,StateSequence)`, for the case of delays/durations, or the predicate `computeFiniteComput ation(InitialState,MaxStepNum,StateSequence)`, for the case of concentration levels. The result is a sequence of states starting from `InitialState` and performing at most `MaxStepNum` steps in the case of concentration levels, while in the delays/durations case the interpreter will give the user the choice to perform also a possibly infinite computation. A computation can also stop when the state gets empty. The predicate `reactionSet/1` defines the list of reactions for the case of delays/durations, while `reactionsQ/1` defines the list of reactions for the case of concentration levels and they have to be redefined for each example to be studied.

6.1   Case Study: Controlling the Differentiation in Th-Cell

The immune system is composed by various cell types, including antigen cells and B and T lymphocytes. Among the latter, T cells can be further sub-classified into T helper 1 (*Th1*) or T helper 2 (*Th2*) cells, originating from a common precursor *Th0*. A complex gene network regulates the differentiation of *Th0* cells. Studying the molecular mechanisms of this differentiation process is relevant since enhanced *Th1* and *Th2* responses may cause autoimmune and allergic diseases, respectively.

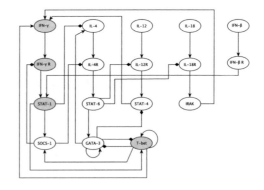

Fig. 4. Graphical representation of the Boolean network

---
[2] https://www3.diism.unisi.it/~falaschi/ReactionSystemsQuantities.

In [16] a Boolean network model of such a regulatory process has been conceived from the large amount of molecular data available in the literature. The network includes 17 nodes regulating the differentiation of the *Th0* precursor [1,19]. The Boolean network is depicted in Fig. 4. Details about the Boolean update functions are in Fig. 5.

In [5] the authors translated the Boolean network into a closed RS (a RS without environment) that used different entities to model different levels for the gray nodes in Fig. 4. The RS can reproduce the dynamics of the update functions in Fig. 5. However, this translation was not very natural because it to introduce the artificial concept of *valid state* to avoid entites representing different levels of the same node to be present at the same time.

Here, we rephrase such a translation using the new linear functions framework presented in Sect. 5. We express different concentrations levels with concentration values $\{1, 2\}$, where 1 stands for medium and 2 for high. For example, the state $\{STAT\text{-}1(1),\ T\text{-}bet\ (2)\}$ states that we have a medium concentration of $STAT$-1 and a high concentration of T-bet.

$$
\begin{aligned}
IFN\text{-}\gamma\text{-}m(t+1) &= (STAT\text{-}4(t) \wedge \neg IRAK(t) \wedge \neg T\text{-}bet\text{-}h(t)) \vee (T\text{-}bet\text{-}m(t) \wedge (\neg STAT\text{-}4(t) \vee \neg IRAK(t))) \\
IFN\text{-}\gamma\text{-}h(t+1) &= (STAT\text{-}4(t) \wedge IRAK(t)) \vee T\text{-}bet\text{-}h(t) \\
IL\text{-}4(t+1) &= GATA\text{-}3(t) \wedge \neg STAT\text{-}1\text{-}m(t) \wedge \neg STAT\text{-}1\text{-}h(t) \\
IFN\text{-}\gamma R\text{-}m(t+1) &= IFN\text{-}\gamma(t)\text{-}m \vee (IFN\text{-}\gamma\text{-}h(t) \wedge SOCS\text{-}1(t)) \\
IFN\text{-}\gamma R\text{-}h(t+1) &= IFN\text{-}\gamma\text{-}h(t) \wedge \neg SOCS\text{-}1(t) \\
IL\text{-}4R(t+1) &= IL\text{-}4(t) \wedge \neg SOCS\text{-}1(t) \\
IL\text{-}12R(t+1) &= IL\text{-}12(t) \wedge \neg STAT\text{-}6(t) \\
IL\text{-}18R(t+1) &= IL\text{-}18(t) \wedge \neg STAT\text{-}6(t) \\
IFN\text{-}\beta R(t+1) &= IFN\text{-}\beta(t) \\
STAT\text{-}1\text{-}m(t+1) &= (IFN\text{-}\beta R(t) \wedge \neg IFN\text{-}\gamma R\text{-}h(t)) \vee IFN\text{-}\gamma R\text{-}m(t) \\
STAT\text{-}1\text{-}h(t+1) &= IFN\text{-}\gamma R\text{-}h(t) \\
GATA\text{-}3(t+1) &= (STAT\text{-}6(t) \vee GATA\text{-}3(t)) \wedge \neg (T\text{-}bet\text{-}h(t) \vee T\text{-}bet\text{-}m(t)) \\
SOCS\text{-}1(t+1) &= T\text{-}bet\text{-}m(t) \vee T\text{-}bet\text{-}h(t) \vee STAT\text{-}1\text{-}m(t) \vee STAT\text{-}1\text{-}h(t) \\
IRAK(t+1) &= IL\text{-}18R(t) \\
STAT\text{-}4(t+1) &= IL\text{-}12R(t) \wedge \neg GATA\text{-}3(t) \\
STAT\text{-}6(t+1) &= IL\text{-}4R(t) \\
T\text{-}bet\text{-}m(t+1) &= (STAT\text{-}1\text{-}m(t) \vee T\text{-}bet\text{-}m(t)) \wedge \neg (STAT\text{-}1\text{-}h(t) \vee T\text{-}bet\text{-}h(t) \vee GATA\text{-}3(t)) \\
T\text{-}bet\text{-}h(t+1) &= (STAT\text{-}1\text{-}h(t) \vee T\text{-}bet\text{-}h(t)) \wedge \neg GATA\text{-}3(t)
\end{aligned}
$$

Fig. 5. Boolean functions modelling the differentiation of Th cells

We advocate that using the linear functions framework has many advantages compared to the modelling approach adopted in [5]. For example, in [5] reaction $(\{GATA\text{-}3\}, \{STAT\text{-}1\text{-}h,\ STAT\text{-}1\text{-}m\ \}, \{IL\text{-}4\})$ was used to describe the production of IL-4, which is inhibited by $STAT$-1 in high or medium concentration. This required to include two inhibitors in the rule (distinguished by the final -h and -m, respectively). Instead, in the our new framework based on linear functions the same event can be modeled by the following simpler reaction: $\{GATA\text{-}3(1)\}, \{STAT\text{-}1(1)\}, \{IL\text{-}4(1)\})$. It is worth noticing that such reaction is enabled in any state containing $GATA$-3, but that does not containing $STAT$-1 at any level, as desired. Similarly, in [5] the production of $SOCS$-1 when T-bet at any level is present was modelled by reactions ({ *T-bet*-h },{},{*SOCS*-1}) and ({ *T-bet*-m},{},{*SOCS*-1}). In the linear functions framework the production of *SOCS*-1 can be expressed by the single reaction ({*T-bet*(1)},{},{*SOCS*-1(1)}). Even more interestingly, in [5] two reactions ({*IFN-γ*R-m},{},{*STAT*-1-m}), ({*IFN-γ*R-h},{},{*STAT*-1-h}) were introduced to express the fact that *IFN-γ*R at some level produces *STAT*-1 at the same level. In the linear functions framework one reaction suffices: ({ *IFN-γ*R (x+1)},{},{*STAT*-1(x+1)}).

As a result we obtain 26 reactions, available online[3], that model the system described in Fig. 5. Our prototype implementation allows us to compute the LTS. We performed two in silico experiments that show some paths leading to *Th1* differentation (note that the presence of T-bet is the marker of the differentiation of the cell into *Th1* form). In the first one, the evolution is driven by the up-regulation of *IFN-γ* that is expressed at the maximal level at the initial state: {*IFN-γ*(2)}. After 6 steps we reach the stable state {*IFN-γ*(2), *IFN-γ*R(2), *STAT*-1(2), *T-bet*(2)} that shows that the differentiation towards *Th1* cell is successfully accomplished. The second experiment is driven by the initial expression of both *IL*-12 and *IL*-18. The initial state in this case is {*IL*-12, *IL*-18}, and, after 9 steps, the system reaches the stable state. Of course, all intermediate states can be inspected, and this allows us to observe that the second experiment reaches soon a high level of *IFN-γ*, and then the execution continues as in the first experiment.

7   Conclusions and Related Work

The model of RSs is qualitative as there is no direct representation of the number of molecules involved in biochemical reactions as well as of rate parameters influencing the frequency of reactions. In [17] the authors introduce an extension with discrete concentrations allowing for quantitative modelling. They demonstrate that although RSs with discrete concentrations are semantically equivalent to the original qualitative RSs, they provide much more succinct representations in terms of the number of molecules being used. They then define the problem of reachability for RSs with discrete concentrations, and provide its suitable encoding in satisfiability modulo theory, together with a verification method (bounded model checking) for reachability properties. Experimental results show that verifying RSs with discrete concentrations instead of the corresponding basic RS is more efficient. A crucial feature of a RS is that (unless introduced from outside the system) an entity from the current state will belong also to the next state only if it is in the product set of an enabled reaction. In other words, an entity vanishes unless it is sustained by a reaction. In [7] it is introduced an extension where such a property is mitigated, indeed they provide each entity x with a duration $d(x)$, which guarantees that x will last through at least $d(x)$ consecutive states. The authors demonstrate that duration/decay is a result of an interaction with a "structured environment", and they also investigate fundamental properties of state sequences of reaction systems with duration". Each

---

[3] https://www3.diism.unisi.it/~falaschi/reactionsConcentrationLevels.txt.

of these enhancements of the RS framework requires complex changes in the syntax and semantics of the original framework and they cannot easily be combined together. Our semantic framework for RSs is more flexible, since it allows us to define extensions by simply playing with the defined SOS rules. We have shown this possibility by defining extensions with reaction delays and durations in Sect. 4, and with concentration levels in Sect. 5. Also adapting our prototype tool form RS execution was made easier by the SOS formalization. It is worth noting that these and other extensions can be combined and integrated in our framework by following the same approach.

As future work we plan to exploit our framework to deepen the study of quantitative extensions of RSs. In particular, we will continue the investigation of the extensions we introduced in this paper also by evaluating their applicability to case studies of biochemical pathways and gene regulation networks. This will be done without violating the discrete and abstract nature of RSs. Moreover, the availability of a formal semantics will allow us to study and apply formal analysis techniques aimed at assessing dynamical properties of the modelled biological systems. Finally, we plan to investigate the applicability of abstract interpretation techniques [10,11] to study properties of classes of reaction systems by exploiting under- and over-approximations of current states, which is particularly convenient when quantitative information is present in the system.

References

1. Agnello, D., et al.: Cytokines and transcription factors that regulate t helper cell differentiation: new players and new insights. J. Clin. Immunol. **23**(3), 147–161 (2003). https://doi.org/10.1023/A:1023381027062
2. Azimi, S.: Steady states of constrained reaction systems. Theor. Comput. Sci. **701**(C), 20–26 (2017). https://doi.org/10.1016/j.tcs.2017.03.047
3. Azimi, S., Iancu, B., Petre, I.: Reaction system models for the heat shock response. Fundam. Informaticae **131**(3–4), 299–312 (2014). https://doi.org/10.3233/FI-2014-1016
4. Barbuti, R., Gori, R., Levi, F., Milazzo, P.: Investigating dynamic causalities in reaction systems. Theor. Comput. Sci. **623**, 114–145 (2016). https://doi.org/10.1016/j.tcs.2015.11.041
5. Barbuti, R., Gori, R., Milazzo, P.: Encoding Boolean networks into reaction systems for investigating causal dependencies in gene regulation. Theor. Comput. Sci. (2021). https://doi.org/10.1016/j.tcs.2020.07.031
6. Brijder, R., Ehrenfeucht, A., Main, M., Rozenberg, G.: A tour of reaction systems. Int. J. Found. Comput. Sci. **22**(07), 1499–1517 (2011). https://doi.org/10.1142/S0129054111008842
7. Brijder, R., Ehrenfeucht, A., Rozenberg, G.: Reaction systems with duration. In: Kelemen, J., Kelemenová, A. (eds.) Computation, Cooperation, and Life. LNCS, vol. 6610, pp. 191–202. Springer, Heidelberg (2011). https://doi.org/10.1007/978-3-642-20000-7_16
8. Brodo, L., Bruni, R., Falaschi, M.: A logical and graphical framework for reaction systems. Theor. Comput. Sci. **875**, 1–27 (2021). https://doi.org/10.1016/j.tcs.2021.03.024

9. Corolli, L., Maj, C., Marinia, F., Besozzi, D., Mauri, G.: An excursion in reaction systems: from computer science to biology. Theor. Comput. Sci. **454**, 95–108 (2012). https://doi.org/10.1016/j.tcs.2012.04.003
10. Cousot, P.: Principles of Abstract Interpretation. MIT Press, Cambridge (2021)
11. Cousot, P., Cousot, R.: Abstract interpretation: a unified lattice model for static analysis of programs by construction or approximation of fixpoints. In: Proceedings of the ACM POPL 1977, pp. 238–252. ACM (1977). https://doi.org/10.1145/512950.512973
12. Ehrenfeucht, A., Main, M.G., Rozenberg, G.: Combinatorics of life and death for reaction systems. Int. J. Found. Comput. Sci. **21**(3), 345–356 (2010). https://doi.org/10.1142/S0129054110007295
13. Ehrenfeucht, A., Main, M.G., Rozenberg, G.: Functions defined by reaction systems. Int. J. Found. Comput. Sci. **22**(1), 167–178 (2011). https://doi.org/10.1142/S0129054111007927
14. Ehrenfeucht, A., Rozenberg, G.: Reaction systems. Fundam. Informaticae **75**(1–4), 263–280 (2007)
15. Hillston, J.: A compositional approach to performance modelling. Ph.D. Thesis, University of Edinburgh, UK (1994)
16. Mendoza, L.: A network model for the control of the differentiation process in th cells. Biosystems **84**(2), 101–114 (2006). https://doi.org/10.1016/j.biosystems.2005.10.004
17. Męski, A., Koutny, M., Penczek, W.: Towards quantitative verification of reaction systems. In: Amos, M., Condon, A. (eds.) Unconventional Computation and Natural Computation, UCNC 2016. LNCS, vol. 9726, pp. 142–154. Springer, Cham (2016). https://doi.org/10.1007/978-3-319-41312-9_12
18. Milner, R. (ed.): A Calculus of Communicating Systems. LNCS, vol. 92. Springer, Heidelberg (1980). https://doi.org/10.1007/3-540-10235-3
19. Murphy, K.M., Reiner, S.L.: Decision making in the immune system: the lineage decisions of helper t cells. Nat. Rev. Immunol. **2**, 933–944 (2002). https://doi.org/10.1038/nri954
20. Okubo, F., Yokomori, T.: The computational capability of chemical reaction automata. Nat. Comput. **15**(2), 215–224 (2015). https://doi.org/10.1007/s11047-015-9504-7
21. Pardini, G., Barbuti, R., Maggiolo-Schettini, A., Milazzo, P., Tini, S.: Compositional semantics and behavioural equivalences for reaction systems with restriction. Theor. Comput. Sci. **551**, 1–21 (2014). https://doi.org/10.1016/j.tcs.2014.04.010
22. Plotkin, G.D.: An operational semantics for CSP. In: Bjørner, D. (ed.) Proceedings of the IFIP Working Conference on Formal Description of Programming Concepts-II, Garmisch-Partenkirchen, pp. 199–226. North-Holland (1982)
23. Plotkin, G.D.: A structural approach to operational semantics. J. Log. Algebraic Methods Program. **60–61**, 17–139 (2004). https://doi.org/10.1016/j.jlap.2004.05.001
24. Villasana, M., Radunskaya, A.: A delay differential equation model for tumor growth. J. Math. Biol. **47**(3), 270–294 (2003). https://doi.org/10.1007/s00285-003-0211-0

Dynamic Heuristic Set Selection for Cross-Domain Selection Hyper-heuristics

Ahmed Hassan$^{(\boxtimes)}$ and Nelishia Pillay(iD)

Department of Computer Science, University of Pretoria, Pretoria, South Africa
ahmedhassan@aims.ac.za, nelishia.pillay@up.ac.za

Abstract. Selection hyper-heuristics have proven to be effective in solving various real-world problems. Hyper-heuristics differ from traditional heuristic approaches in that they explore a heuristic space rather than a solution space. These techniques select constructive or perturbative heuristics to construct a solution or improve an existing solution respectively. Previous work has shown that the set of problem-specific heuristics made available to the hyper-heuristic for selection has an impact on the performance of the hyper-heuristic. Hence, there have been initiatives to determine the appropriate set of heuristics that the hyper-heuristic can select from. However, there has not been much research done in this area. Furthermore, previous work has focused on determining a set of heuristics that is used throughout the lifespan of the hyper-heuristic with no change to this set during the application of the hyper-heuristic. This paper investigates dynamic heuristic set selection (DHSS) which applies dominance to select the set of heuristics at different points during the lifespan of a selection hyper-heuristic. The DHSS approach was evaluated on the benchmark set for the CHeSC cross-domain hyper-heuristic challenge. DHSS was found to improve the performance of the best performing hyper-heuristic for this challenge.

Keywords: Dynamic heuristic set selection · Selection perturbative hyper-heuristics · Cross-domain hyper-heuristics

1 Introduction

Selection hyper-heuristics explore a heuristic space to choose a constructive or perturbative heuristic at each point to create or improve a solution respectively [14]. Previous work has shown that the performance of selection hyper-heuristics is affected by the set of the constructive or perturbative heuristics that the hyper-heuristic selects from [11,15,16]. Generally, the entire set of heuristics available for the problem domain, which we refer to as the *universal set*, is used by the hyper-heuristic, increasing the search space to be explored which may harm the performance of the hyper-heuristic. We hypothesize that the set of heuristics available to the hyper-heuristic to perform selections from, the *active set*, should be different at different points in the lifespan of the hyper-heuristic.

© Springer Nature Switzerland AG 2021
C. C. Aranha et al. (Eds.): TPNC 2021, LNCS 13082, pp. 33–44, 2021.
https://doi.org/10.1007/978-3-030-90425-8_3

We investigate this hypothesis for cross-domain selection perturbative hyper-heuristics.

Previous work in this area has shown the effectiveness of determining an active set at the beginning of the lifespan of the hyper-heuristic, instead of the hyper-heuristic selecting from the entire set of heuristics. In this case, the active set is static, i.e. the same set is used throughout the lifespan of the hyper-heuristic. In [16], fitness landscape measures are used to estimate the performance of heuristics which are subsequently ranked using non-parametric tests. Several active sets are generated by considering the heuristics according to their ranks. In [6], two ranking methods are used to filter out poor and redundant heuristics: the first method is based on the gain (heuristic performance) and correlation, and the second method is based on the z-test. However, there does not appear to be any research into using different active heuristic sets at different points in the lifespan of a hyper-heuristic. We refer to this as *dynamic heuristic set selection* (DHSS).

This study examines DHSS for cross-domain selection perturbative hyper-heuristics. A dominance approach is used to select the active heuristic set at different points in the lifespan of the hyper-heuristic. The approach is evaluated using the CHeSC challenge benchmark set. It is implemented with the best-performing hyper-heuristic for the challenged, namely, FS-ILS [1]. DHSS was found to improve the cross-domain performance of FS-ILS and FS-ILS together with DHSS outperformed state-of-the-art approaches for the challenge. Hence, the contributions of the research presented in this study are:

1. A DHSS approach for selection hyper-heuristics.
2. An investigation of DHSS for cross-domain hyper-heuristics.

The rest of this paper is organized as follows. The following section defines terms used in the paper in the context of the research presented. Section 3 provides an overview of cross-domain hyper-heuristics and the CHeSC challenge. In Sect. 4, the approach is described in detail. Section 5 presents the experimental setup. Section 6 discusses the performance of DHSS. Section 7 concludes the paper and presents directions for future research.

2 Terminology

This section presents some terminology that will be used in the paper. The *lifespan* of a selection perturbative hyper-heuristic is a single application of the hyper-heuristic comprised of several iterations. The *universal set* refers to the set of all heuristics for a problem domain. The *active set* is the set of heuristics used at a particular point in the lifespan of a hyper-heuristic. This is often a subset of the universal set. A *heuristic* refers to a problem-specific heuristic that is an element of the universal set. A *phase* is a part of the lifespan in which the active set remains fixed and a *phase length* is the number of iterations executed during the phase. A *duration* of a heuristic is the total execution time in milliseconds used by the heuristic.

3   Cross-Domain Hyper-heuristics

Cross-domain hyper-heuristics aim at achieving a higher level of generality by solving problems across different problem domains [14]. In this case, hyper-heuristics aim to perform well across all the problem domains rather than producing good results for some problems and poor results for others. The CHeSC challenge aimed at promoting research in cross-domain hyper-heuristics [5]. The challenge required a selection perturbative hyper-heuristic to be developed such that it performed well in six problem domains, namely, the boolean satisfiability problem, one-dimensional bin packing problem, personnel scheduling problem, permutation flow shop problem, traveling salesman problem, and vehicle routing problem. The HyFlex framework [13] was developed for the challenge. HyFlex provides an implementation of heuristics, methods for creating initial solutions, and methods for calculating the objective value for each problem domain. For each problem domain, HyFlex provides four types of heuristics. *Mutational heuristics* perturb a given solution at random. *Ruin and recreate heuristics* destroy and rebuild some parts of a given solution. *Local search heuristics* improve a given solution. *Crossover heuristics* recombine parts from two solutions to produce a new solution.

The selection perturbative hyper-heuristics competing in the challenge were ranked and the ranks are added to obtain the overall score. In cases where there is a tie, the corresponding points to the relevant positions are added together and shared equally among all algorithms that tie. The median of the objective value over 31 independent runs is used to rank the competing hyper-heuristics using Formula 1 in which the top 8 methods (hyper-heuristics) receive a score of 10, 8, 6, 5, 4, 3, 2, 1 points respectively and the rest receive no points. The cross-domain score is calculated by adding up all problem-specific scores and the winner is the hyper-heuristic with the highest cross-domain score. A time limit of 600 s is used for each run and a benchmark program is provided to estimate the time limit on different machines so that it matches the 600 s on the standard machine used in CHeSC.

Research into producing a competitive selection perturbative hyper-heuristic is still ongoing. We provide an overview of the six best performing hyper-heuristics at the time of writing this paper. Fair share iterated local search (FS-ILS) [1] is a hyper-heuristic that outperforms the winner of CHeSC. FS-ILS uses fitness proportionate selection to select heuristics based on their ability to generate accepted solutions and the duration taken to achieve that. FS-ILS employs a randomized local search step and a Metropolis acceptance criterion. A restart strategy is also incorporated to re-initialize the search if it stagnates for very long. In this paper, we show that the cross-domain performance of FS-ILS can be improved if DHSS is used.

The winner of the CHeSC challenge was adapHH [12] which maintains a subset of heuristics for each phase. adapHH selects a heuristic based on a probability calculated by considering the best improvements and the duration of the heuristic. Relay hybridization is used to identify heuristics that work well in pairs. An adaptive threshold acceptance criterion is used to accept worsening solutions.

VNS [8] based on a variable neighborhood search working on a population of solutions was placed second in the challenge. VNS shakes a solution, which is chosen from the population by a tournament selection. A perturbative heuristic is applied to the solution to achieve this. Then, a local search is applied to the solution. A tabu list is used to keep track of worsening perturbative heuristics. The worse solutions in the population are replaced by better solutions generated by the local search.

The (ML) approach developed by Mathieu Larose [4,13] was placed third in the challenge. ML is an iterated local search hyper-heuristics that employ reinforcement learning to select heuristics. The method consists of a diversification step performed by perturbative heuristics, an intensification step performed by local search heuristics, and an acceptance criterion that accepts worsening moves only if the incumbent solution has not improved for several iterations.

Pearl Hunter (PHunter) [3], placed fourth in the challenge, is an iterated local search that mimics pearl hunters. The search involves two steps: diversification (surface moves) and intensification (dive). PHunter can try more than one diversification move if the search is trapped in a "buoy". PHunter performs two types of local searches: "snorkeling", i.e. short sequences of local search and "scuba dive", i.e. intensive local search.

Evolutionary programming hyper-heuristic (EPH) [10], placed fifth in the challenge, co-evolves two populations: a population of sequences of heuristics that are applied to the solutions in the other population. Each sequence consists of one or two perturbative heuristics (mutational and ruin-recreate heuristics) followed by all local search heuristics.

4   Dynamic Heuristic Set Selection (DHSS) Approach

This section describes the DHSS approach. DHSS uses dominance [2] to determine the active set at each point in the lifespan of the hyper-heuristic. The DHSS is described in Algorithm 1.

Algorithm 1 requires the universal set and search-status information from the hyper-heuristic which includes information such as the current iteration, the elapsed time, and the current solution value. The algorithm starts by initializing the active set to include all heuristics as in line 1. In line 2, the set of permanently excluded heuristics is initialized as an empty set. In line 3, the performance history is initialized for all heuristics. The history keeps track of information about each heuristic such as the percentage improvement, the percentage non-improvement, and duration.

The update condition (in line 4 of Algorithm 1) is checked at the start of every iteration of the hyper-heuristic and the active set is updated if it fails to improve the best solution for N_{fail} iterations where N_{fail} is determined as $pf \times wait_{max}$ where $wait_{max}$ is the maximum number of iterations that have elapsed between two consecutive updates of the best solution and pf is a *patience factor* which controls how fast/slow the updates occur. Further, we observed in some instances, $wait_{max}$ can grow very large if the search stagnates for a long

---

Algorithm 1. DHSS.

Require: universal set $\mathcal{U}$, search-status information $\mathcal{I}$ from the hyper-heuristic
1: $S \leftarrow \mathcal{U}$ // Initialize the active set S
2: $S' \leftarrow \emptyset$ // Initialize the set of permanently removed heuristics
3: Initialize performance history $\mathcal{P}$ for all heuristics
4: **if** canUpdate($\mathcal{I}$) **then**
5:      **if** canRemove($\mathcal{I}$) **then**
6:          $S' \leftarrow$ remove(S, S', $\mathcal{P}$, $\mathcal{I}$)
7:      **end if**
8:      $S \leftarrow$ update(S, S', $\mathcal{P}$)
9:      return S
10: **else**
11:     return S
12: **end if**

---

period before the best solution is updated. For this reason, we set a maximum phase length ($phase_{max}$) to ensure that the active set is still updated even if $wait_{max}$ grows very large.

The removal condition (line 5 of Algorithm 1) is $T_{elp}/T_{max} > R_{excl}$ where T_{elp} is the time elapsed since the start of the hyper-heuristic, T_{max} is the total computational time, and R_{excl} is a parameter that controls how soon the permanent heuristic removal is performed. The heuristic removal is executed once during the search at the first update that happens after T_{elp}/T_{max} exceeds R_{excl}.

The removal criterion (line 6 of Algorithm 1) excludes a heuristic permanently only if it has poor *individual performance* and poor *group performance*. The individual performance of a heuristic is measured from its own history such as the ratio between the number of improvements made by the heuristic and the total number of times the heuristic is used. The group performance considers the performance of a heuristic relative to all heuristics such as the ratio between the percentage improvement made by the heuristic and the total percentage improvement made by all heuristics. The individual performance of a heuristic h_i is defined as follows:

$$f_i^{ind} = \alpha_1 \underbrace{\frac{n_i^+}{n_i^+ + n_i^-} - \alpha_2 \frac{n_i^-}{n_i^+ + n_i^-}}_{frequency} + \alpha_3 \underbrace{\frac{\Delta_i^+}{\Delta_i^+ + \Delta_i^-} - \alpha_4 \frac{\Delta_i^-}{\Delta_i^+ + \Delta_i^-}}_{amount} \qquad (1)$$

where n_i^+ and n_i^- denote the number of improvements and non-improvements respectively made by heuristic i; Δ_i^+ and Δ_i^- denote the amount of percentage improvement and percentage non-improvement respectively made by heuristic i; and $0 \leq \alpha_i \leq 1, i = 1, 2, 3, 4$ are weights which are used to give some terms more importance. The group performance of a heuristic h_i is defined as follows:

$$f_i^{gp} = \beta_1 \underbrace{\frac{n_i^+}{\sum\limits_{k=1}^{H} n_k^+} - \beta_2 \frac{n_i^-}{\sum\limits_{k=1}^{H} n_k^-}}_{\text{frequency}} + \beta_3 \underbrace{\frac{\Delta_i^+}{\sum\limits_{k=1}^{H} \Delta_k^+} - \beta_4 \frac{\Delta_i^-}{\sum\limits_{k=1}^{H} \Delta_k^-}}_{\text{amount}} \qquad (2)$$

where all symbols are as defined in Eq. (1), H is the total number of heuristics, and $0 \le \beta_i \le 1, i = 1, 2, 3, 4$ are weights.

We calculate the averages $\bar{f}_{ind}$ and $\bar{f}_{gp}$ for all values of f_i^{ind} and f_i^{gp} respectively. A heuristic h_i is removed permanently only if $f_i^{ind} < \bar{f}_{ind}$ and $f_i^{gp} < \bar{f}_{gp}$. This criterion removes only the worst-performing heuristics that worsen the current solution, do this often, and fail to compensate for this degenerating behavior by producing significant improvements.

The update criterion (line 8 of Algorithm 1) decides which heuristics should be included in the active set. The update criterion utilizes a *measure*, which evaluates the heuristic performance, to determine suitable heuristics for the next phase. In this study, we measure the heuristic performance by the *frequency of improvements* which favors heuristics that generates more improving moves. The active set is updated such that it includes all *dominant* heuristics and excludes all *dominated* heuristics. A heuristic h_i dominates a heuristic h_k if $v_i > v_k$ and $t_i < t_k$ where v_j is the value of heuristic j as determined by the measure and t_j is the duration of heuristic h_j in milliseconds. This criterion does not exclude a worse heuristic unless it has a longer duration than a better heuristic. The reason for considering the duration is that very slow heuristics are not desirable in general.

5   Experimental Setup

This section describes the experimental setup in terms of parameter tuning and technical specifications.

5.1   Parameter Settings

The parameters involved in DHSS are summarized in Table 1. Manual tuning via trial and error is used to determine the best values of these parameters. For each parameter, several values are tried and the best value is chosen as reported in Table 2. For the α_i and β_i ($i = 1, 2, 3, 4$) in Eq. (1) and Eq. (2), we tried 3 different configurations:

1. Assign more importance to improvements than non-improvements. In this case, we have $\alpha_1 = \alpha_3 = \beta_1 = \beta_3 = 1.0$ and $\alpha_2 = \alpha_4 = \beta_2 = \beta_4 = 0.5$. This configuration achieves the best results.
2. Assign more importance to non-improvements than improvements. In this case, we have $\alpha_1 = \alpha_3 = \beta_1 = \beta_3 = 0.5$ and $\alpha_2 = \alpha_4 = \beta_2 = \beta_4 = 1.0$.
3. No distinction. In this case, we have $\alpha_i = \beta_i = 1.0$ for $i = 1, 2, 3, 4$.

Table 1. Summary of the parameters of DHSS.

Parameter	Brief description
$phase_{max}$	Phase length
pf	Patience factor which controls how fast the updates occur
R_{excl}	Controls when to start removing some heuristics permanently
$\alpha1, \beta_1$	Weight of frequency of improvement in Eq. (1) and Eq. (2) respectively
$\alpha2, \beta_2$	Weight of frequency of non-improvement in Eq. (1) and Eq. (2) respectively
$\alpha3, \beta_3$	Weight of percentage improvement in Eq. (1) and Eq. (2) respectively
$\alpha4, \beta_4$	Weight of percentage non-improvement in Eq. (1) and Eq. (2) respectively

Table 2. Parameters values

Parameter	Tried values	Best value
$phase_{max}$	1, 100, 1000	100
pf	0.2, 0.5, 1.0	0.5
R_{excl}	0.2, 0.5, 1.5	0.2

We observed that no parameter setting leads to best performance in all problem domains. The best values reported in Table 2 generate the best overall cross-domain performance. The most influential parameters are $phase_{max}$ and R_{excl}. A drop in the performance by about 40% is observed between the best and worst value for $phase_{max}$. The worst value of R_{excl} deteriorates the performance by about 24%.

As part of the parameter tuning, we also tried resetting the active set to include all heuristics that are not permanently removed when the search stagnates for a number of iterations exceeding $wait_{max}$. However, this did not lead to improvement; hence it is not included in Algorithm 1.

5.2   Technical Specification

The experiments are executed in Java 8 on the Lengau Cluster of the Center for High-Performance Computing, South Africa. The cluster's operating system is CentOS 7.0. We used two compute nodes to perform 31 independent runs in parallel. Each node has 24 Intel Xeon CPUs (2.6 GHz) and is connected with FDR 56 GHz InfiniBand. The total RAM used is 2 GB.

Table 3. FS-ILS compared to the top five hyper-heuristics from the CHeSC Challenge.

Method	FS-ILS [1]	adapHH [12]	VNS [8]	ML [13]	PHunter [3]	EPH [10]
Overall score	178.10	161.68	117.18	111.0	81.60	73.60

Table 4. The cross-domain performance of DHSS compared to the top five methods of CHeSC and FS-ILS.

Method	DHSS	FS-ILS [1]	adapHH [12]	VNS [8]	ML [13]	PHunter [3]	EPH [10]
Overall score	178.75	148.85	142.10	98.10	94.75	71.60	59.10

6   Results and Discussion

This section presents and discusses the results of DHSS across the six domains of HyFlex used in the CHeSC Challenge.

6.1   Performance Verification

In this section, we verify that FS-ILS is the state-of-the-art hyper-heuristic. We respect all CHeSC competition conditions. The results are presented in Table 3. The cross-domain scores are calculated by adding up all individual scores for each problem domain as described in Sect. 3. Higher scores correspond to better performance. For brevity, the table shows the top five methods. The cross-domain score of FS-ILS confirms that it is indeed the best hyper-heuristic.

6.2   DHSS and CHeSC Contestants

We evaluate the performance of DHSS with respect to the contestants of CHeSC and FS-ILS across the six domains of HyFlex. We respected all the CHeSC rules. The methods are scored using the CHeSC scoring system described in Sect. 3 where higher scores indicate better performance. Although we compare DHSS to all contestants of CHeSC, for brevity, Table 4 presents the cross-domain scores for the top five methods, and the best hyper-heuristic for CHeSC (FS-ILS). DHSS improves the cross-domain performance of FS-ILS and achieves the highest cross-domain score. This demonstrates the effectiveness of utilizing DHSS to manage the heuristic set for hyper-heuristics.

The per-domain results are presented in Fig. 1 which shows that DHSS achieves the best performance in 3 domains (SAT, PFS, TSP) and the second-best performance in VRP. No hyper-heuristic dominates all other hyper-heuristics in all domains. For each hyper-heuristic, there is at least one domain that poses a challenge to it. For example, DHSS performs poorly in BP, adapHH performs poorly in PS, and EPH is unable to score any points in SAT; hence, its column is not shown in the figure. In all domains, DHSS performs consistently better than FS-ILS except for BP.

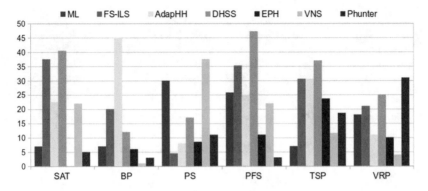

Fig. 1. The performance of DHSS compared to the top five methods of CHeSC and FS-ILS.

6.3   Analysis of DHSS

In the previous section, we empirically demonstrated the effectiveness of DHSS in improving the cross-domain performance of the best hyper-heuristic for the CHeSC cross-domain challenge. In this section, we investigate the reasons behind the performance gain. We hypothesize that when DHSS is used, good heuristics will have a larger share of the total computational time, hence utilized more; and poor heuristics will have a smaller share of the total computational time, hence utilized less. To this end, we calculated how much percentage improvement each heuristic contributed to the total percentage improvement. This is measured by the ratio between the percentage improvement made by the heuristic and the total percentage improvement made by all heuristics. We also calculate the share of each heuristic in the total computational time which is measured by the ratio between the total computational time used by the heuristic and the total computational time allocated to the hyper-heuristic.

We present the results for two representative domains (SAT and PFS) due to space limitations. From each domain, one instance is chosen arbitrarily. We measured the shares of each heuristic in the total percentage improvement and computational time, as explained above, using 10 runs with different seeds.

The results for SAT are presented in Figs. 2 and 3 for the perturbative and local search heuristics respectively. In these figures and all following figures, we use the HyFlex convention in identifying each heuristic by a unique number. These heuristics are described in [9] and [17] for SAT and PFS respectively. The most effective perturbative heuristics for SAT are h2, h3, and h5 which receive collectively an increase of 11% in computational time in DHSS compared to FS-ILS. Furthermore, the local search heuristic h8 is much less effective than h7 since it leads to a marginal improvement. However, in FS-ILS, h8 has an unnecessarily large share of 41% of the total computational time whereas, DHSS restricts the computational time of h8 to 11% of the total computational time.

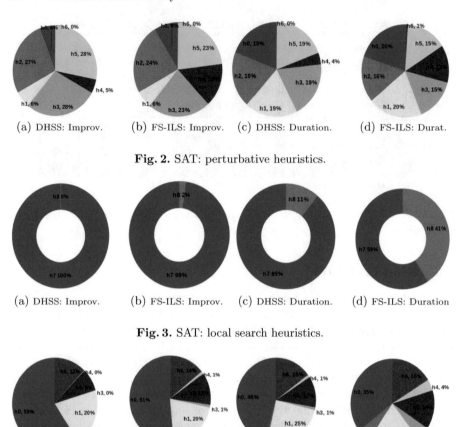

(a) DHSS: Improv.   (b) FS-ILS: Improv.   (c) DHSS: Duration.   (d) FS-ILS: Durat.

Fig. 2. SAT: perturbative heuristics.

(a) DHSS: Improv.   (b) FS-ILS: Improv.   (c) DHSS: Duration.   (d) FS-ILS: Duration

Fig. 3. SAT: local search heuristics.

(a) DHSS: Improv.   (b) FS-ILS: Improv.   (c) DHSS: Duration.   (d) FS-ILS: Duration

Fig. 4. PFS: perturbative heuristics.

The results of the analysis for PFS are presented in Figs. 4 and 5 for perturbative and local search heuristics respectively. The most effective perturbative heuristics are h0, h1, h5, and h6. The heuristic h0 is responsible for more than 50% of the total improvement. The computational time of h0 is 11% larger in DHSS compared to FS-ILS. Similarly, h1 has an increase of 5% in computational time in DHSS compared to FS-ILS. On the other hand, h5 and h6 have slightly more computational time in FS-ILS compared to DHSS. In general, good heuristics receive collectively an increase in the computational time of 13% in DHSS compared to FS-ILS. Moreover, poor heuristics (h2, h3, and h4) collectively use 14% of the total computational time in FS-ILS despite collectively contributing by 2% to the total percentage improvement. In DHSS, the collective computational time of h2, h3, and h4 is restricted to only 3% of the total computational time. For local search heuristics, both DHSS and FS-ILS perform well.

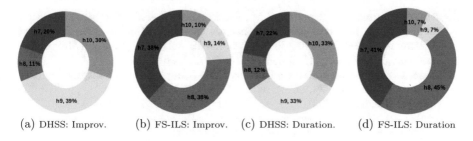

(a) DHSS: Improv.   (b) FS-ILS: Improv.   (c) DHSS: Duration.   (d) FS-ILS: Duration

Fig. 5. PFS: local search heuristics.

7   Conclusion and Future Work

In this paper, we solved the problem of determining an adequate heuristic set for selection hyper-heuristics dynamically such that the heuristic set changes at different points during the lifespan of the hyper-heuristic. We integrated the proposed approach (DHSS) into the state-of-the-art hyper-heuristic (FS-ILS) and evaluated it across six problem domains. The results indicated that the performance of FS-ILS was improved when the DHSS was used. We carried out an analysis to discover the reasons behind the performance gain. It was found that when the DHSS was used, good heuristics had a larger share of the total computational time, hence utilized more; and poor heuristics had a smaller share of the total computational time, hence utilized less. DHSS can be used with any hyper-heuristics since it does not rely on any specific feature of FS-ILS. We developed an open-source Java library to enable fast development and prototyping of DHSS [7].

In the future, we will consider automating the design of dynamic heuristic set selection approaches. It is also interesting to consider DHSS as a design decision in a wider-scope approach that aims at automating the design of multiple aspects of the hyper-heuristics simultaneously.

Acknowledgments. This work is funded as part of the Multichoice Research Chair in Machine Learning at the University of Pretoria, South Africa. This work is based on the research supported wholly/in part by the National Research Foundation of South Africa (Grant Numbers 46712). Opinions expressed and conclusions arrived at, are those of the author and are not necessarily to be attributed to the NRF. This work is run on the Lengau Cluster of the Center for High Performance Computing, South Africa.

References

1. Adriaensen, S., Brys, T., Nowé, A.: Fair-share ILS: a simple state-of-the-art iterated local search hyperheuristic. In: Proceedings of the 2014 Annual Conference on Genetic and Evolutionary Computation, pp. 1303–1310 (2014)

2. Alvarez-Benitez, J.E., Everson, R.M., Fieldsend, J.E.: A MOPSO algorithm based exclusively on pareto dominance concepts. In: Coello Coello, C.A., Hernández Aguirre, A., Zitzler, E. (eds.) EMO 2005. LNCS, vol. 3410, pp. 459–473. Springer, Heidelberg (2005). https://doi.org/10.1007/978-3-540-31880-4_32

3. Chan, C.Y., Xue, F., Ip, W.H., Cheung, C.F.: A hyper-heuristic inspired by pearl hunting. In: Hamadi, Y., Schoenauer, M. (eds.) LION 2012. LNCS, pp. 349–353. Springer, Heidelberg (2012). https://doi.org/10.1007/978-3-642-34413-8_26

4. Drake, J.H., Kheiri, A., Özcan, E., Burke, E.K.: Recent advances in selection hyper-heuristics. Eur. J. Oper. Res. **285**(2), 405–428 (2020)

5. Burke, E.K., et al.: The cross-domain heuristic search challenge – an international research competition. In: Coello, C.A.C. (ed.) LION 2011. LNCS, vol. 6683, pp. 631–634. Springer, Heidelberg (2011). https://doi.org/10.1007/978-3-642-25566-3_49

6. Gutierrez-Rodríguez, A.E., et al.: Applying automatic heuristic-filtering to improve hyper-heuristic performance. In: 2017 IEEE Congress on Evolutionary Computation (CEC), pp. 2638–2644. IEEE (2017)

7. Hassan, A., Pillay, N.: Java library for dynamic heuristic set selection, September 2021. https://github.com/Al-Madina/Dynamic-Heuristic-Sets

8. Hsiao, P.C., Chiang, T.C., Fu, L.C.: A VNS-based hyper-heuristic with adaptive computational budget of local search. In: 2012 IEEE Congress on Evolutionary Computation, pp. 1–8. IEEE (2012)

9. Hyde, M., Ochoa, G., Vázquez-Rodríguez, J.A., Curtois, T.: A hyflex module for the max-sat problem. University of Nottingham, Technical report, pp. 3–6 (2011)

10. Meignan, D.: An evolutionary programming hyper-heuristic with co-evolution for CHeSC11. In: The 53rd Annual Conference of the UK Operational Research Society (OR53), vol. 3 (2011)

11. Mısır, M., Verbeeck, K., De Causmaecker, P., Vanden Berghe, G.: The effect of the set of low-level heuristics on the performance of selection hyper-heuristics. In: Coello, C.A.C., Cutello, V., Deb, K., Forrest, S., Nicosia, G., Pavone, M. (eds.) PPSN 2012. LNCS, vol. 7492, pp. 408–417. Springer, Heidelberg (2012). https://doi.org/10.1007/978-3-642-32964-7_41

12. Mısır, M., Verbeeck, K., De Causmaecker, P., Vanden Berghe, G.: An intelligent hyper-heuristic framework for CHeSC 2011. In: Hamadi, Y., Schoenauer, M. (eds.) LION 2012. LNCS, pp. 461–466. Springer, Heidelberg (2012). https://doi.org/10.1007/978-3-642-34413-8_45

13. Ochoa, G., et al.: HyFlex: a benchmark framework for cross-domain heuristic search. In: Hao, J.-K., Middendorf, M. (eds.) EvoCOP 2012. LNCS, vol. 7245, pp. 136–147. Springer, Heidelberg (2012). https://doi.org/10.1007/978-3-642-29124-1_12

14. Pillay, N., Qu, R.: Hyper-Heuristics: Theory and Applications. Natural Computing Series, Springer, Heidelberg (2018). https://doi.org/10.1007/978-3-319-96514-7

15. Pillay, N.: A review of hyper-heuristics for educational timetabling. Ann. Oper. Res. **239**(1), 3–38 (2016)

16. Soria-Alcaraz, J.A., Ochoa, G., Sotelo-Figeroa, M.A., Burke, E.K.: A methodology for determining an effective subset of heuristics in selection hyper-heuristics. Eur. J. Oper. Res. **260**(3), 972–983 (2017)

17. Vázquez-Rodrıguez, J.A., Ochoa, G., Curtois, T., Hyde, M.: A hyflex module for the permutation flow shop problem. School of Computer Science, University of Nottingham, Technical report (2009)

Deep Learning and Transfer Learning

On the Transfer Learning of Genetic Programming Classification Algorithms

Thambo Nyathi[✉][iD] and Nelishia Pillay[iD]

Department of Computer Science, University of Pretoria,
Lynnwood Road, Pretoria 0002, South Africa
{t.nyathi,nelishia.pillay}@up.ac.za

Abstract. Data classification is a real-world problem that is encountered daily in various problem domains. Genetic programming (GP) has proved to be one of the most versatile algorithms leading to its popularity as a classification algorithm. However, due to its large number of parameters, the manual design process of GP is considered to be a time consuming tedious task. As a result, there have been initiatives by the machine learning community to automate the design of GP classification algorithms. In this paper, we propose the transfer of the design knowledge gained from the automated design of GP classification algorithms from a specific source domain and apply it to design GP classification algorithms for a target domain. The results of the experiments demonstrate that the proposed approach is capable of evolving classifiers that achieve results that are competitive when compared to automated designed classifiers and better than manually tuned parameter classifiers. To the best of our knowledge, this is the first study that examines transfer learning in automated design. The proposed approach is shown to achieve positive transfer.

Keywords: Data classification · Transfer learning · Genetic programming · Automated design

1 Introduction

Genetic programming (GP) [4] has proven to be one of the most commonly used evolutionary algorithms [13]. The representation used by GP provides it with flexibility which allows GP to be used in a wide range of problem domains. One such popular domain is data classification, where GP is used as a classification algorithm. However, like most evolutionary algorithms GP is highly parameterised and thus its effectiveness depends on its configuration. Configuration of GP for data classification is an algorithm design process that requires certain design decisions to be made. Traditionally, the algorithm design process has been carried out manually using parameter tuning through a trial and error approach. Manual design and configuration have been shown to be time consuming, tedious and non-trivial processes that lead to human bias and mistakes [3]. Furthermore,

© Springer Nature Switzerland AG 2021
C. C. Aranha et al. (Eds.): TPNC 2021, LNCS 13082, pp. 47–58, 2021.
https://doi.org/10.1007/978-3-030-90425-8_4

Montero and Riff [5] argue that although not always possible knowledge of both the problem domain and classification algorithm design is essential. As a result of these constraints, the Machine Learning research community has taken an initiative towards automated design of machine learning and search algorithms [10]. A number of studies can be found in the literature that propose the automated design of evolutionary algorithms [10]. In [7] the feasibility of automating the design of genetic programming classification algorithms is shown.

Classification is a problem specific task that requires a classification algorithm to first learn patterns from training data instances resulting in a classifier whose effectiveness can be evaluated on unseen test data. Classification is therefore a supervised learning task that requires labelled data. However, there are instances where labelled data may not be available or there is a time constraint limiting the training of a classification algorithm. In such cases there is a desire for machine learning techniques to emulate how human beings learn i.e. through knowledge transfer. Humans can generalise from past experience in a different problem domain. The ability to transfer knowledge is known as *transfer learning* [9]. In this study, the effectiveness of transfer learning in the automated design of genetic programming classification algorithms for data classification is evaluated. Two genetic programming classification algorithm designs evolved by automated design on source data are transferred and applied to different target data. The performance of the classifiers evolved by the transferred designed classification algorithms is compared to the performance of manually tuned and automated designed algorithms. The results of the experiments show that on average the classifiers evolved through transferred learning performed better than manually tuned and automated designed classification algorithms. To the best of our knowledge this is the first work that proposes the transfer of designs for the automated design of genetic programming classification algorithms

The rest of the paper is structured as follows: Sect. 2 introduces transfer learning while in Sect. 3 GP and classification are discussed. The proposed approach is introduced in Sect. 4 and the experimental setup is presented in Sect. 5. Section 6 outlines the results and the conclusion and future work are presented in Sect. 7.

2   Transfer Learning

Human beings can effortlessly apply transfer learning in a wide range of day-to-day tasks. However, in computing specifically, in machine learning transfer learning is not a seamless task. A number of formal definitions of transfer learning have been proposed in the literature [9,12]. The definition provided in [9] is receiving wide acceptance and is summarised as follows: Provided with a problem in a source domain denoted as D_S the learning task for this source domain is denoted as T_S. Transfer learning aims to improve the output of function $f(t)$ applied on the target domain D_T using knowledge gained from D_S and T_S. The objective is to improve the functionality of T_T where $D_S \neq D_T$ and $T_S \neq T_T$. According to Pan and Yang [9] three distinct attributes need to be considered when it comes to transfer learning in machine learning. Firstly, *what to transfer*, this involves identifying knowledge in a particular algorithm that is not domain (task) specific and that can be transferred to a similar or different domain (task). Secondly, *when to transfer* i.e. identifying when the knowledge transfer can take place. For example in a classification problem knowledge transfer may take place before or after the training stage. Finally, *how to transfer*. This involves the implementation of the transfer, i.e. is an algorithm transferred as is or an adaptive implementation where the algorithm is altered so as to be applied to the target. Torrey and Shavlink [12] argue for 3 characteristics that should be used to measure success in transfer learning. These are listed as follows: i) at a minimum using the transferred knowledge algorithm should not result in a worse result than using an algorithm (ignorant) that is built for the target domain. ii) applying the transferred knowledge algorithm should result in a time constraint benefit. iii) the final performance of a transferred knowledge algorithm should result in positive transfer i.e. improvement on an ignorant algorithm. When applying the transfer learning results in a worse performance than an ignorant algorithm it is known as *negative* learning. A comprehensive survey of transfer learning is presented in [9].

3   Genetic Programming and Classification

Genetic programming has proven to be a very popular classification algorithm, particularly, in evolving classifiers for binary classification tasks [1]. When configured as a classification algorithm genetic programming utilises supervised learning. As a first step GP initially creates a random population of classifiers. The fitness (usually accuracy) of each individual (classifier) is evaluated on a specified problem. If the desired fitness or some other termination criteria is not met then genetic operators are applied to selected individuals, to evolve a population for the next generation. From generation to generation the classifiers are improved as they are evolved gradually until the specified termination criteria is met. Depending on the type of classifiers required the syntax tree representation can be configured to represent arithmetic, logical or decision tree classifiers.

As stated in Sect. 1 there are several design issues that relate to the design decisions that affect the manual design of genetic programming classification algorithms these are outlined generally in [10] and in detail in [6]. As a result of the problems associated with manual design, there have been a number of initiatives proposing the automated design of such algorithms. The next section discusses the automated design method relevant to this study.

3.1   Automated Design of GP Classification Algorithm

In [7] we proposed the automation of the design of genetic programming classification algorithms. In that study a genetic algorithm [2] and grammatical evolution [8] were used to automatically configure three design decisions of GP classification algorithms namely parameter values, genetic operators and control flow. These are described as follows:

Parameter Values. a) *classifier type* - this specifies the type of classifier i.e. arithmetic, logic or decision tree, b) *population size* - specifies the population size of classifiers, c) *tree generation method* - this could be the full, grow or ramped-half-and-half method, d) *initial tree depth* - is a numeric value that specifies initial tree depth, e) *selection method* - the chosen selection method between fitness proportionate and tournament selection, f) *selection size* - numeric value if the chosen method is tournament selection this specifies the size of the tournament, g) *reproduction rates* - this specifies the application rates of the genetic operators, h) *mutation type* - specifies the type of mutation selected between grow and shrink mutation, i) *mutation depth* - numeric value specifying the limit of the depth of the offspring from the mutation operator, j) *fitness type* - specifies one of five options obtained from the confusion matrix, namely, i) accuracy ii) $f_{measure}$ iii) weighted$_{accuracy}$ iv) weighted$_{rand}$ and v) true positive rate. The details of the implementation of these fitness functions are provided in [7]. k) *Number of generations* - specifies the number of generations before the algorithm terminates.

Genetic Operators. This design decision specifies the combination of genetic operators applied.

Control Flow. This determines the flow of the algorithm i.e. the order in which processes occur.

A detailed presentation of the automated design of genetic programming classification algorithms for data classification is outlined in [7]. The GP design components and the range of possible values are summarized in Table 1.

4   Proposed Approach

The aim of this study is to evaluate the effectiveness of transfer learning in the design of the genetic programming classification algorithm. To achieve this aim we formulate an approach that conforms to the definition presented in Sect. 2.

$$f(s) = T_S(C_S.D_S) \tag{1}$$

Where $f(s)$ is the accuracy obtained on the source domain, T_S is the design or configuration of the GP classification algorithm while C_S are classifiers evolved by the T_S design and applied on source data denoted by D_S.

We propose an approach that transfers the best GP classification algorithm design evolved by an automated design algorithm for a specific source domain to an automated design algorithm for a target domain. Figure 1 depicts the overall architecture of the proposed approach with respect to the study presented in this paper. The top half of the diagram illustrates the flow of automated design processes specifically, AutoGE, from **A** to **B** to **C** and finally **D**. **A** to **D** follow the processes described in Sect. 3.1. The source domain for this specific study was the security domain. This is further explained in Sect. 5. The lower half of the

Table 1. Design components.

Param. description	Range of possible values
Representation	0 - arithmetic, 1 - logical, 2 - decision
Population size	100, 200, 300
Tree generation	0 - full, 1 - grow, 2 - ramped half-and-half
Initial tree depth	2–15 (decision tree 2–8)
Max offspring depth	2–15
Selection method	0 - fitness proportionate, 1 - tournament selection
Selection size	2–10
Reproduction rates	0–100 crossover (mutation = 100-crossover)
Mutation type	0 - grow mutation, 1 - shrink mutation
Max mutation depth	2–6
Control flow	0 - fixed, 1 - random
Operator pool	0–6
Fitness type	0–4
Number of generations	50, 100, 200

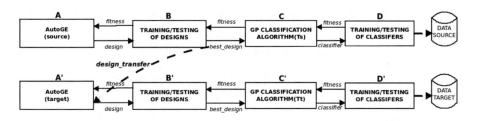

Fig. 1. Transfer learning

diagram illustrates the target automated design specifically, AutoGE henceforth referred to as AutoGE$_T$. The processes in the AutoGE$_T$ algorithm are similar to the source algorithm and are specified from **A'** to **B'** to **C'** and finally to **D'**. The output of process **B** is the *best design* which is the knowledge transferred and used as the initial population of designs for process **A'** to **D'**. In short we reuse the best GP design from a number of possible designs learned by an AutoGE in **A** on a security problem domain and we use the designs in AutoGE$_T$ to evolve classifiers to classify a target domain. The target domains are described in Sect. 5. The proposed approach also provides answers to the questions posed in Sect. 2 as follows:

1. **What to transfer?**
   What is transferred is the design of the GP classification algorithm T_S evolved by AutoGE as presented in Sect. 3.1. The transferred source design T_S becomes the target T_T and is used to evolve classifiers C_T which are trained and tested on the target data D_T.

$$f(t) = T_T(C_T.D_T) \tag{2}$$

2. **When to transfer?**
   Transfer occurs at the design stage before algorithm execution.
3. **How to transfer?**
   The designs are adopted as they are from the source domain and applied to the target domain.

   The next section describes the experimental set up for the proposed approach.

5   Experiment Setup

To evaluate the effectiveness of the proposed approach we obtained two AutoGE evolved source designs T_{S1} and T_{S2} from a prior study [7]. The source data of the security problem domain was obtained from the NSL-KDD 99+20% [11] security dataset. These designs were selected for two reasons, firstly, they achieved near optimal accuracy values T_{S1} 98% and T_{S2} 99% on the security domain. Secondly, they specified different classifier types i.e. T_{S1} arithmetic classifiers and T_{S2} logical classifiers. The target data was obtained from the UCI machine learning data repository and is outlined in Table 2. In a similar approach as in [7] each dataset was randomly split into 70% for training and 30% for testing.

Table 2. Summary of selected data sets.

Dataset	# Attributes	# Numeric	# Nominal	# Instances
Australian credit data	14	8	6	690
Appendicitis	7	7	0	106
Breast cancer (Ljubljana)	9	0	9	277
Cylinder band	19	19	0	365
Diabetes (pima)	8	8	0	768
German credit data	20	7	13	1000
Heart disease	13	13	0	270
Hepatitis	19	19	0	80
Liver disease (Bupa)	7	7	0	345
Mushroom	22	0	22	5644
Tictactoe	9	0	9	958

The following subsections describe the experiments conducted.

5.1   AutoGE$_T$

As previously stated in a prior study AutoGE was found to be effective at configuring GP classification algorithms for binary problems. Therefore, it was of interest to evaluate the effectiveness of transferring the best performing designs for a different problem domain and using the designs as the initial population of AutoGE as opposed to randomly generating an initial population of designs. In this experiment we configured the AutoGE with T_{S1} (termed T_{T1} after transfer) and T_{S2} (termed T_{T2} after transfer) as the initial population. The specific values of the transferred designs T_{T1} and T_{T2} are listed in Table 4 in column 2 and 3 respectively. The automated design with grammatical evolution is termed AutoGE$_T$. Table 3 presents the AutoGE$_T$ parameters.

Table 3. AutoGE$_T$ settings.

Parameter	Value
Population size	20
Initial population	$10\,T_{S1}$, $10\,T_{S2}$
Selection method	Tournament (size 4)
Single point crossover rate	85%
Bit mutation rate	5%
Elitism	10%
Fitness function	Accuracy
Individual size	14–16
Wrapping	Yes
Maximum generations	30
Independent runs	30

5.2 Manually Tuned Transferred Designs

In order to evaluate the effectiveness of the automated design approach's ability to optimise the transferred designs baseline experiments using manually tuned transferred designs were also carried out. Certain parameters of the transferred designs were manually tuned while the other parameter values were kept at their transferred values. The designs were then used to configure GP classification algorithms for data classification and applied to the same target data as AutoGE$_T$. The experiments are described as follows:

Reuse (T_{T1} and T_{T2}). In this set of experiments the designs T_{S1} and T_{S2} obtained from [7] were reused without any alterations to evolve classifiers which were applied to each of the 11 datasets. The configurations are outlined in Table 4. The first column refers to the design decision of the GP configuration while the column 2 and column 3 are the values of the design decisions contained in the designs T_{T1} and T_{T2} respectively.

Table 4. Designs.

Design decision	T_{T1}	T_{T2}	T_{GO1}	T_{GO2}	T_{CF1}	T_{CF2}
Tree type	0	1	0	1	0	1
Population size	100	200	100	200	100	200
Tree generation method	1	0	1	0	1	0
Initial tree depth	5	8	5	8	5	8
Max offspring depth	6	9	6	9	6	9
Selection method	1	0	1	0	1	0
Selection size	4	0	4	0	4	0
Crossover rate	24	58	**80**	**80**	24	58
Mutation type	0	0	0	0	0	0
Mutation depth	4	6	4	6	4	6
Control flow	0	1	0	1	**1**	**0**
Operator combination	0	1	0	1	0	1
Fitness function	0	0	0	0	0	0
Number of generations	100	100	100	100	100	100

Genetic Operators (T_{GO1} and T_{GO2}). This set of experiments involved manually tuning only the application rates of the genetic operators design decision on each of the transferred designs. The application rates of the genetic operators namely, crossover and mutation of crossover 80% and mutation 20% were found to be the most suitable. Table 4 column 4 (T_{GO1}) and column 5 (T_{GO2}) illustrates the parameter values used for this set of experiments.

Control Flow (T_{CF1} and T_{CF2}). In this set of experiments the control flow of T_{S1} and T_{S2} were manually tuned to the values illustrated in column 6 and column 7 of Table 4 these are denoted as T_{CF1} and T_{CF2}.

    In addition to the AutoGE$_T$ experiment six experiments (six designs) were conducted each design was used to configure a GP classification algorithm and applied to each of the data sets outlined in Table 2. For each design and on each dataset 30 independent runs were performed with the best training classifier being applied to a corresponding test set.

6   Results and Discussion

This section presents the results and provides a discussion of the results. The results are outlined in the Tables as follows:

- Column 1, Datasets.
- Column 2, AutoGE, these are results of the automated designed algorithms applied to the target data without any knowledge transfer, what is referred in [12] as an ignorant algorithm due to its lack of transferred knowledge.

- Column 3, TunedParam these are the results of the manually tuned parameter algorithms with no transferred knowledge.
- Column 4, $AutoGE_T$, the results presented in this column are of the automated designed approach with transferred designs.
- Column 5 (T_{T1}) and column 6 (T_{T2}) present the accuracy results of GP classifiers after directly applying the transferred designs to configure GP.
- Column 7 (T_{GO1}) and column 8 (T_{GO2}) present the accuracy results of classifiers after configuring GP with partially manually optimised transferred designs. The optimisation was manually performed on the genetic operators only.
- Column 9 (T_{CF1}) and column 10 (T_{CF2}) presents the accuracy results of applying the algorithms after partially manually optimising the control flow.

Results presented in columns 1 and 2 of Tables 5 and 6 are obtained from [7].

6.1 Training Results

Table 5 shows the training results obtained from applying the seven algorithms to the datasets. The first column is a listing of the datasets while the second column up to the ninth column are a listing of the training results of classifiers evolved by the specified algorithms. Each row refers to a dataset while a column is the applied algorithm. The best training accuracy ± standard deviation over 30 runs is presented for each algorithm.

From the results the $AutoGE$ algorithm achieved a better accuracy on 5 datasets while the $TunedPara$ algorithms trained better on 2 datasets. The T_{GO1} and T_{CF1} each trained well on 1 dataset and tied on 1. The following algorithms T_{T1}, T_{GO2} and T_{CF2} tied on 1 dataset each. On average across all datasets the T_{GO1} trained better than all other algorithms with a training average of 89%. The $AutoGE_T$ algorithm had the least training average of 83% across all datasets. This may be due to the fact that there is a lack of diversity during the early stages of the search since at initialisation the search starts from 2 specific points (2 transferred designs) on the search space as opposed to a randomly created initial population.

Table 5. Training results.

Dataset	AutoGE	TunedPara	$AutoGE_T$	T_{T1}	T_{T2}	T_{GO1}	T_{GO2}	T_{CF1}	T_{CF2}
Australia credit	0.91 ± 0.01	0.91 ± 0.01	0.86 ± 0.03	0.89 ± 0.03	0.87 ± 0.03	0.90 ± 0.01	0.89 ± 0.03	$\mathbf{0.93 \pm 0.01}$	0.87 ± 0.03
Appendix	0.95 ± 0.02	$\mathbf{0.97 \pm 0.02}$	0.93 ± 0.01	0.93 ± 0.01	0.83 ± 0.08	0.95 ± 0.02	0.85 ± 0.08	0.90 ± 0.01	0.84 ± 0.08
Breast cancer	$\mathbf{0.99 \pm 0.01}$	0.98 ± 0.01	0.97 ± 0.01	0.97 ± 0.01	0.93 ± 0.02	0.98 ± 0.02	0.97 ± 0.02	0.98 ± 0.01	0.93 ± 0.02
Cylinder band	0.80 ± 0.04	0.77 ± 0.01	0.80 ± 0.04	$\mathbf{0.82 \pm 0.01}$	0.77 ± 0.01	0.81 ± 0.01	0.77 ± 0.01	0.81 ± 0.01	0.77 ± 0.01
Diabetes	0.74 ± 0.01	$\mathbf{0.78 \pm 0.07}$	0.66 ± 0.04	0.71 ± 0.04	0.68 ± 0.04	0.73 ± 0.04	0.74 ± 0.04	0.72 ± 0.04	0.68 ± 0.04
German credit	$\mathbf{0.86 \pm 0.01}$	0.76 ± 0.06	0.75 ± 0.06	0.77 ± 0.06	0.74 ± 0.06	0.77 ± 0.01	0.77 ± 0.01	0.76 ± 0.03	0.74 ± 0.01
Heart disease	$\mathbf{0.95 \pm 0.01}$	0.94 ± 0.01	0.90 ± 0.04	0.89 ± 0.04	0.86 ± 0.04	0.89 ± 0.04	0.84 ± 0.04	0.88 ± 0.04	0.86 ± 0.04
Hepatitis	0.98 ± 0.04	0.98 ± 0.03	0.92 ± 0.04	0.95 ± 0.04	0.91 ± 0.04	$\mathbf{1.00 \pm 0.00}$	0.89 ± 0.04	0.96 ± 0.05	0.91 ± 0.04
Liver	0.76 ± 0.01	0.80 ± 0.01	0.70 ± 0.01	0.77 ± 0.01	$\mathbf{1.00 \pm 0.00}$	$\mathbf{1.00 \pm 0.00}$	$\mathbf{1.00 \pm 0.00}$	$\mathbf{1.00 \pm 0.00}$	$\mathbf{1.00 \pm 0.08}$
Mushroom	$\mathbf{0.88 \pm 0.00}$	0.86 ± 0.00	0.85 ± 0.00	0.86 ± 0.00	0.85 ± 0.00	0.86 ± 0.00	0.86 ± 0.00	0.86 ± 0.01	0.85 ± 0.00
Tictactoe	$\mathbf{0.99 \pm 0.01}$	0.87 ± 0.07	0.85 ± 0.07	0.87 ± 0.07	0.76 ± 0.07	0.89 ± 0.07	0.84 ± 0.07	0.85 ± 0.03	0.76 ± 0.07
Average	$\mathbf{0.89 \pm 0.02}$	0.87 ± 0.06	0.83 ± 0.03	0.86 ± 0.03	0.84 ± 0.04	$\mathbf{0.89 \pm 0.02}$	0.86 ± 0.03	0.88 ± 0.02	0.84 ± 0.04

6.2   Test Results

Table 6 presents the test results obtained from applying the best training classifier from each algorithm on unseen test data. The classifiers evolved by automated design, $AutoGE$ tested better on 2 datasets and tied on two datasets. The $AutoGE_T$ evolved classifiers performed better on four datasets and tied on one. The T_{T1} algorithm performed well on one dataset and tied on one. The $TunedPara$ algorithm and T_{TGO1} tested well on one dataset each while T_{T2} and T_{CF1} algorithms tied on the same dataset. The T_{TGO2} algorithm tied on two datasets. On average across all datasets the $AutoGE_T$ algorithm tested better than all the other algorithms with an average accuracy of 83%. This is followed by the $AutoGE$, T_{GO1} and T_{CF1} algorithms which all had an individual test average classification of 81%. T_{T1} and T_{CF2} both individually tested at 80% while T_{T2} tested at 79%. Notably, the manually parameter tuned algorithms have the least testing average accuracy of 77%. To evaluate the statistical significance of the differences in performance between the algorithms the two-tailed pairwise Wilcoxon's signed-rank test was used [14]. All the tests were performed at the $\alpha = 0.05$ significance level. In [7] it was shown that the differences in performance between the $AutoGE$ algorithm and the manually tuned algorithms were significant. The differences in performance between the $AutoGE$ and the $AutoGE_T$ was found not to be statistically significant. The differences in performance between $AutoGE_T$ and the manually tuned algorithms were found to be statistically significant. The performances of T_{GO1} and T_{CF1} were found to be not statistically significant. Notably, the difference between the $AutoGE_T$ and the worst performing design altered configurations T_{T2} and T_{GO2} were also found to be not statistically significant. The statistical significance in performance between the manually tuned parameter algorithms and the best performing transferred algorithms were found to be not significant. This is also the case between the manually tuned algorithms and the worst performing transferred algorithms. Considering test results it is evident that the $AutoGE_T$ has the best overall performance of the *learned* algorithms. The $AutoGE_T$ performed numerically better than the $AutoGE$ algorithm. Depending on the type of classification error minimisation is sometimes an important attribute in classification. This points to the potential of transfer learning in GP classification algorithms.

Table 6. Testing results.

Dataset	AutoGE	TunedPara	$AutoGE_{Trf}$	T_{T1}	T_{T2}	T_{GO1}	T_{GO2}	T_{CF1}	T_{CF2}
Australia credit	0.86 ± 0.01	0.85 ± 0.01	$\mathbf{0.88 \pm 0.01}$	0.85 ± 0.01	0.85 ± 0.01	0.82 ± 0.01	0.86 ± 0.01	0.86 ± 0.01	0.85 ± 0.01
Appendix	$\mathbf{0.94 \pm 0.01}$	0.85 ± 0.01	0.91 ± 0.01	0.93 ± 0.01	0.78 ± 0.10	$\mathbf{0.94 \pm 0.01}$	0.78 ± 0.08	0.82 ± 0.05	0.81 ± 0.05
Breast cancer	$\mathbf{0.98 \pm 0.02}$	0.97 ± 0.02	$\mathbf{0.98 \pm 0.02}$	0.97 ± 0.02	0.93 ± 0.03	0.97 ± 0.02	0.94 ± 0.02	0.97 ± 0.02	0.94 ± 0.02
Cylinder band	0.74 ± 0.04	0.69 ± 0.01	0.84 ± 0.01	$\mathbf{0.84 \pm 0.01}$	0.80 ± 0.01	0.80 ± 0.01	0.75 ± 0.01	0.83 ± 0.01	0.80 ± 0.01
Diabetes	0.60 ± 0.04	$\mathbf{0.75 \pm 0.07}$	0.73 ± 0.07	0.71 ± 0.07	0.71 ± 0.07	0.68 ± 0.07	0.70 ± 0.07	0.68 ± 0.07	0.68 ± 0.07
German credit	0.66 ± 0.01	0.65 ± 0.01	$\mathbf{0.67 \pm 0.01}$	0.66 ± 0.01	0.64 ± 0.01	0.65 ± 0.01	0.66 ± 0.01	0.63 ± 0.01	0.63 ± 0.01
Heart disease	0.81 ± 0.08	0.77 ± 0.02	$\mathbf{0.89 \pm 0.04}$	0.80 ± 0.08	0.75 ± 0.09	0.80 ± 0.08	0.73 ± 0.05	0.75 ± 0.09	0.75 ± 0.09
Hepatitis	0.88 ± 0.01	0.75 ± 0.06	$\mathbf{0.95 \pm 0.01}$	0.79 ± 0.06	0.79 ± 0.06	0.75 ± 0.06	0.75 ± 0.06	0.88 ± 0.06	0.79 ± 0.06
Liver	0.65 ± 0.01	0.64 ± 0.06	0.71 ± 0.05	0.74 ± 0.06	$\mathbf{1.00 \pm 0.00}$	0.99 ± 0.01	$\mathbf{1.00 \pm 0.00}$	$\mathbf{1.00 \pm 0.00}$	0.99 ± 0.01
Mushroom	$\mathbf{0.81 \pm 0.00}$	0.78 ± 0.00	$\mathbf{0.81 \pm 0.00}$	$\mathbf{0.81 \pm 0.00}$	0.80 ± 0.00	0.72 ± 0.00	$\mathbf{0.81 \pm 0.00}$	0.78 ± 0.02	0.78 ± 0.00
Tictactoe	$\mathbf{0.98 \pm 0.01}$	0.76 ± 0.01	0.80 ± 0.01	0.73 ± 0.01	0.69 ± 0.01	0.79 ± 0.01	0.77 ± 0.01	0.69 ± 0.01	0.80 ± 0.01
Average	0.81 ± 0.02	0.77 ± 0.03	$\mathbf{0.83 \pm 0.02}$	0.80 ± 0.04	0.79 ± 0.04	0.81 ± 0.03	0.79 ± 0.03	0.81 ± 0.03	0.80 ± 0.03

7   Conclusion and Future Work

This paper proposed a knowledge transfer approach for the automated design of GP classification algorithms. A GP classification algorithm design effective on the security problem source domain was transferred and applied to multiple problem target domains. On average across all datasets the proposed approach results in the evolution of classifiers that perform numerically better than the algorithms with no knowledge transfer. The best performing algorithm is the automated design algorithm that uses the transferred designs as the starting points of the search. This approach produces results that are statistically better than manual parameter tuning. This results in an overall reduced design time as transferred designs can be used as they are or with alterations. As outlined in [7] the design time for manual parameter tuning can range from a number of days to weeks. The results of the experiments showed that the transfer learned classifiers are statistically comparable to the automated designed classifiers. It is therefore, safe to conclude that there is no negative learning that occurs which shows the potential of the proposed approach. Results from the study indicate that GP classification algorithms classifiers evolved by both automated design with knowledge transfer and manual design with partial knowledge transfer are able to achieve improved accuracy results. The results of the study conforms to the transfer learning requirements stipulated by Torrey and Shavlink [12]. The proposed approach does not result in evolving classifiers that perform worse than algorithms with no transferred knowledge and there is certainly no negative learning that takes place.

Future work will attempt to evaluate if there is any correlation between the source domain and the target domain. If there is a specific design from a source domain that is suitable for a specific target domain. We will also investigate the impact of transfer learning on the computational effort during algorithm execution. Also of interest would an investigation of the dynamics of the changes in the fitness landscape between search spaces evolved by transfer learning compared to parameter tuned fitness landscapes.

References

1. Espejo, P.G., Romero, C., Ventura, S., Hervás, C.: Induction of classification rules with grammar-based genetic programming. In: Conference on Machine Intelligence, pp. 596–601 (2005)
2. Goldberg, D.E., Holland, J.H.: Genetic algorithms and machine learning. Mach. Learn. **3**, 95–99 (1988)
3. Hutter, F., Hoos, H.H., Leyton-Brown, K., Stützle, T.: ParamILS: an automatic algorithm configuration framework. J. Artif. Intell. Res. **36**, 267–306 (2009)
4. Koza, J.R.: Concept formation and decision tree induction using the genetic programming paradigm. In: Schwefel, H.-P., Männer, R. (eds.) PPSN 1990. LNCS, vol. 496, pp. 124–128. Springer, Heidelberg (1991). https://doi.org/10.1007/BFb0029742

5. Montero, E., Riff, M.-C.: Towards a method for automatic algorithm configuration: a design evaluation using tuners. In: Bartz-Beielstein, T., Branke, J., Filipič, B., Smith, J. (eds.) PPSN 2014. LNCS, vol. 8672, pp. 90–99. Springer, Cham (2014). https://doi.org/10.1007/978-3-319-10762-2_9

6. Nyathi, T.: Automated design of genetic programming of classification algorithms. Ph.D. thesis (2018)

7. Nyathi, T., Pillay, N.: Comparison of a genetic algorithm to grammatical evolution for automated design of genetic programming classification algorithms. Expert Syst. Appl. **104**, 213–234 (2018)

8. O'Neill, M., Ryan, C.: Grammatical evolution. IEEE Trans. Evol. Comput. **5**(4), 349–358 (2001)

9. Pan, S.J., Yang, Q.: A survey on transfer learning. IEEE Trans. Knowl. Data Eng. **22**(10), 1345–1359 (2009)

10. Pillay, N., Qu, R., Srinivasan, D., Hammer, B., Sorensen, K.: Automated design of machine learning and search algorithms [guest editorial]. IEEE Comput. Intell. Mag. **13**(2), 16–17 (2018)

11. Tavallaee, M., Bagheri, E., Lu, W., Ghorbani, A.A.: A detailed analysis of the KDD cup 99 data set. In: 2009 IEEE Symposium on Computational Intelligence for Security and Defense Applications, pp. 1–6. IEEE (2009)

12. Torrey, L., Shavlik, J.: Transfer learning. In: Handbook of Research on Machine Learning Applications and Trends: Algorithms, Methods, and Techniques, pp. 242–264. IGI Global (2010)

13. Vanneschi, L., Castelli, M., Silva, S.: A survey of semantic methods in genetic programming. Genet. Program Evolvable Mach. **15**(2), 195–214 (2014)

14. Woolson, R.: Wilcoxon signed-rank test. In: Wiley Encyclopedia of Clinical Trials, pp. 1–3 (2007)

On Injecting Entropy-Like Features into Deep Neural Networks for Content Relevance Assessment

Jakub Sido[1,2]([⊠]) [iD], Kamil Ekštein[2] [iD], Ondřej Pražák[1,2] [iD],
and Miloslav Konopík[1,2] [iD]

[1] NTIS – New Technologies for the Information Society, University of West Bohemia,
Pilsen, Czech Republic
{sidoj,ondfa,konopik}@kiv.zcu.cz

[2] Department of Computer Science and Engineering, Faculty of Applied Sciences,
University of West Bohemia, Pilsen, Czech Republic
kekstein@kiv.zcu.cz

Abstract. This paper describes in details an innovative technique of injection of a global (or generally large-scale) quality measure into a deep neural network (DNN) in order to compensate for the tendency of DNNs to found the resulting classification virtually from a superposition of local neighbourhood transformations and projections. We used a *state probability-like feature* as the global quality measure and injected it into a DNN-based classifier deployed in a specific task of determining which parts of a web page are of certain interest for further processing by NLP techniques. Our goal was to decompose web sites of various internet discussion forums to useful content, i.e. the posts of users, and useless content, i.e. forum graphics, menus, banners, advertisements, etc.

Keywords: Deep learning · Entropy · Global information · Content relevance assessment

1 Introduction

Recent advancements in the field of NLP facilitated even more massive ingression of artificial intelligence to various e.g. security or commercial applications that analyze texts posted to a number of internet discussion forums by their users.

There have been numerous applications intensely reliant on crawling[1] and subsequently analyzing the contents of internet discussion forums around (and, of course, social networks of which substantial part of the front-end can be treated as forums, too). The most common are various recommender systems, their arbitrarily complex derivatives, and similar uses. These applications—driven by mostly economical interests—have prevailed so far, however, recently, hand in

---

[1] The process of getting the full (or specifically filtered) content of web pages automatically by the so-called web crawler [14].

© Springer Nature Switzerland AG 2021
C. C. Aranha et al. (Eds.): TPNC 2021, LNCS 13082, pp. 59–68, 2021.
https://doi.org/10.1007/978-3-030-90425-8_5

hand with the escalation of social and political tension in the world, safety and protective applications are emerging with the same needs for training data. A raw crawled web content far from suitable for any further processing since different pieces of information are not clearly differentiated and there is no obvious difference between useful and unwanted content. The Internet contains a wide variety of internet forums with different structure and organization. No matter how different, the forums usually structure the discussions into individual posts. The posts are further split into the author's name (the nick), date of publication and the actual post text (sometimes even more information about the post is provided). Proper processing of the Internet forums requires to correctly extract all the post fields.

Recognizing the posts and their structure is often not sufficient. The web pages are frequently infested with various unwanted content (mainly advertisements). Such content might strongly resemble the useful content for it is the goal of commercial content creators. Naturally, the marketing managers desire that the commercial communication was perceived and understood as a part of the original text (which was specifically sought after by users) as the credited origin imparts considerable trustworthiness and importance to it.

Needless to say, it is highly advisable to get rid of this unwanted content prior to any subsequent algorithmic processing. The authors were initially motivated and driven by the need to distinguish the unwanted content from user posts on various internet discussion forums that were about to be investigated for the presence of potentially interesting content from a law enforcement point of view (e.g. hatred, racism, terrorism, etc.).

Later on, after having carried some experiments, it turned out that the proposed technique is applicable more generally (with such-and-such moderate adjustments) to virtually any situation where there is a demand for an assessment of text relevance.

2   Related Work

Surprisingly, there are only several few relevant works on the topic of forum content assessment and extraction. [3] proposed a technique comparing the structure of two web pages and formalizing it using a grammar. As a result, this technique is able to remove those parts of the pages that are the same (as possibly repeating on each web page several times) and takes as the useful content the stuff that differs.

Such an approach is valid but the procedure works satisfactorily only in the case of forums with entirely static unwanted content. Naturally, modern web advertising technology generates content that is massively customized and dynamically changed for different page visitors and thus making the mentioned method completely useless. The key problem to be dealt with is how to tell that certain (part of a) text is more important than the other. When put the other way round, it is to tell which (part of a) text is unimportant when compared to the wanted content. Using the statements of information theory, we may assume that it is the one that carries less information. Unfortunately, when solely measuring the amount of information (in e.g. bits), we could end up with in-page advertisements that take more bits than the useful content.

However, by examining the ways the advertisements and the commercial content are produced by means of HTML tags and how they are integrated into the web pages [2], there is a way larger number of manners the construction and integration are accomplished compared to the useful content. The useful content is rather regularly structured and built using fewer tags than the unwanted stuff. This observation is particularly obvious in the case of discussion forums where the threads of user posts are structured very specifically and the pattern keeps repeating throughout the whole forum.

Thus, *there is observably lower diversity in the DOM^2 subtrees defining the useful content than in those defining the unwanted content.* In other words, the disorderedness of the useful content is lower compared to the unwanted parts. This consideration naturally leads to the concept of *information entropy* introduced by Claude Shannon [9] as a measure of information production.

Nowadays, due to massive advancements in the field of deep neural networks (DNN), most of the text classification is performed by DNNs [6,8,10,15]. Even though they dispose of an indisputable power to classify (or generally transform) yet a very complex input data, they also suffer from particular drawbacks: Namely, they tend to generalize based upon a rather local neighbourhood in the input data (for instance, the recurrent *long short-term memories* [5] (LSTM) possess only a limited ability to "remember" and thus they are working with a restricted area of the problem space).

Due to that typical property, it does not make much sense to let the DNN learn how to represent the input data entropy as it is a global (or at least non-local) quality. Instead, *entropy or any other non-local feature may be injected into the DNN* from a deterministic numerical computation.

The key concept of the presented approach is to combine DNNs and global metrics in order to distinguish useful and unwanted components of an HTML document as well as to extract and classify the pieces of information of certain interest for further processing.

Unfortunately, due to the tree-like recursive character of the DOM that defines the contents of web pages, it is *rather complicated to express entropy of its parts, e.g. subtrees*, in an easy and straightforward manner (see detailed explanation in Sect. 3). Therefore, a modified *entropy-like measure* based upon relative diversity of the subtrees computed from a frequentistic examination of numbers and types of subtree nodes is proposed in the following section.

---

^{2} DOM = **Document Object Model**, "a cross-platform and language-independent application programming interface that treats an HTML, XHTML, or XML document as a tree structure wherein each node is an object representing a part of the document" [13].

3   Text Relevance Assessment

As outlined in Sect. 2, it is practically impossible to express the entropy using Shannon's definition for subtrees of the DOM representation of a web page.

The Shannon's formula for information entropy computation (Eq. 1) requires to express somehow the probability that a discrete signal sample (as a realization of a random variable) gets a specific value.

$$S = -\sum_i P_i \log P_i \tag{1}$$

This approach is inapplicable to subtrees of the DOM as **the set of all possibly existing subtrees** (the statistical universe of which the size is needed in the denominator) **is infinite** and therefore the frequentistic interpretation of probability collapses – in simple words, it is impossible to compute the probability that a certain subtree (carrying a certain content) appears in the analyzed DOM instance.

Therefore, a modification of Shannon's entropy is needed. Two useful modifications of such kind were shown in [4], however, these were targeted mainly towards discretized continuous signal such as speech (or generally sound). Nonetheless, they served well as basic principles for further consideration and exploration of the problem.

The modification proposed in this work (referred to as *entropy-like measure* hereafter) is grounded in the following statement backed by the above-mentioned observations and considerations: The subtrees with fewer different node types (that are present more often) are labelled with a lower score of the proposed *entropy-like measure*. Equation 2 defines its computation as performed in the experiments discussed below.

$$M_{\text{elf}} = \frac{\sum_{n \in \mathbf{D}_{\text{subtree}}} \log_2 \left(\frac{C_n}{C_{\text{tot}}} \right)}{|\mathbf{D}_{\text{subtree}}|}, \tag{2}$$

where M_{elf} is the numeric value of the *entropy-like measure* for the respective subtree, $\mathbf{D}_{\text{subtree}}$ is the subtree dictionary, i.e. the set of all DOM nodes (HTML tags in fact) present in the respective subtree, C_n is the number of occurrences of a certain node $n, n \in \mathbf{D}_{\text{subtree}}$ in the respective subtree, and C_{tot} is the total number of all nodes in the respective subtree.

Such a definition makes it possible to establish a relation of orderedness on the set of all DOM subtrees constituting the whole DOM of a web page. Thus, the subtrees may be compared whether one or another is more (or less) "chaotic" and therefore carries less (or more) useful content.

The described measure (Eq. 2) is subsequently used to assess all nodes of the DOM of the analyzed web page. Therefore, a value $X_{\text{elf}} \in \langle 0, 1 \rangle$ normalized over the whole web page is computed using Eq. 3:

$$X_{\text{elf}} = \frac{2^{M_{\text{elf}}}}{N_{\text{subtree}}}, \tag{3}$$

where N_{subtree} is the number of nodes in the respective DOM subtree.

The X_{elf} value can be then directly used to mark each subtree in the web page DOM. Examples of a forum web page processed by the above-described algorithm can be seen in Fig. 2 where the DOM nodes (subtrees) are darkened according to the value of X_{elf} for the respective node (subtree).

4   Classifier Architecture and Operation

The experimental classifier was built (and subsequently evaluated) using the **TensorFlow** framework [1] for DNN-based machine learning extended with an in-house overlay facilitating the model design and creation.

The classifier is a typical heterogeneous deep neural network model which uses several neuron layers of different types (recurrent and fully connected), namely *long short-term memories* (LSTM) and *fully connected* (dense) layers.

The architecture of the classifier is depicted in Fig. 1. The input to the classifier is a node of the web page DOM which is to be classified. The output

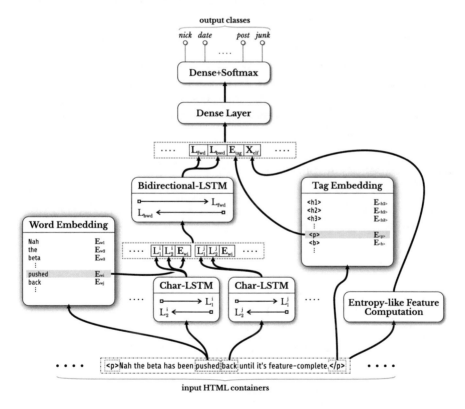

Fig. 1. Architecture of the textual content type classifier with injection of a non-local feature (here specifically the *entropy-like measure*).

(probabilistically interpretable as the output layer is the Softmax one) is a class label $c \in \mathbb{N} \cap \langle 0, 6 \rangle$. The classes are summarized and explained in Table 2.

The DOM node is at first preprocessed: Each container HTML tag (like e.g. <h1>, <h2>, <h3>, <p>, <b>, etc.) is detached from the respective container content and its embedding is concatenated with other features into a vector which is fed into the first dense layer (named "Dense Layer" in Fig. 1) of the classifier.

The container content is processed in a typical text-oriented NLP manner: The text is split into words and each word is transformed by bidirectional character-based LSTM [7] into two scores denoted as L_1^i and L_2^i, respectively. These two scores are then supplemented with the word embedding E_{wi} to create an input word representation for the bidirectional LSTM modeling the whole container content.

The container-level bidirectional LSTM [6,12] produces another two scores, L_{fwd} and L_{bwd} from the forward and the backward LSTM run, respectively. These scores are grouped together with the tag embedding E_{tag} and the *entropy-like feature* X_{elf} obtained from evaluation of the expression defined by Eq. 3 on the respective node (and the whole submerged content).

The sequence of vectors representing the whole subtree $\left(L_{fwd}^{(i)}, L_{bwd}^{(i)}, E_{tag}^{(i)}, X_{elf}^{(i)} \right)_{i=1}^{N_{subtree}}$ is then fed into the dense layer ("Dense Layer" in Fig. 1) whose output is fed into the output layer ("Dense+Softmax" in Fig. 1). The output layer produces probabilistically interpretable classification through activations of its output neurons where each represents one class to which the DOM subtree can be classified (see Table 2).

4.1 Classifier Hyperparameters

The adjustable model settings that were not given in figures in Sect. 4 are summarized in Table 1 below. The hyperparameter names refer to model components with corresponding labels in Fig. 1.

Table 1. Classifier hyperparameters.

Hyperparameter	Value
Char-LSTM size	100
Char embedding dimension	100
Bidirectional-LSTM size	300
Word embedding dimension	300
Tag embedding dimension	100
Dense layer size	256
Dense+Softmax size	256
Learning rate	0.00001
Dropout	0.5

5   Classifier Training

The classifier depicted in Sect. 4 was trained on a special corpus of internet discussion forums collected especially for the sake of training of a content relevance assessment techniques [11]. This dataset contains 79 crawled internet discussion forum sites. From each site (forum), at minimum 501 and at maximum 2317 pages were incorporated into the dataset resulting in 65242 pages in total (on average, 826 pages per forum). The web pages constituting the dataset were manually annotated so that each node of the web page DOM had a certain class label assigned from the set shown in Table 2.

Table 2. Classes of content used for relevance assessment on internet discussion forums. The alternative class names were introduced in the paper describing the dataset used for training.

Class label	Class name	Content type
0	junk (others)	Everything else like ads, banners, navigation, etc.
1	nick	The username used by the post author
2	date	The date of publication of the post
3	post (text)	The post itself, i.e. the published text
4	citation-author	The username of the quoted author
5	citation-date	The date the quoted post was published
6	citation-text	The quote itself, i.e. the quoted text

The dataset was split into two parts for training and evaluation purposes: (i) a *training set* consisting of 85% of the samples, and (ii) a *testing/evaluation set* consisting of the remaining 15% of the samples.

The training was performed as *batch training* with the batch size of 64 samples. The samples from the *training set* were picked randomly (uniform distribution) to build the batches. The whole training took approx. 6 h on the computational grid. After 3 epochs, the loss function drop ceased and the convergence criterion was reached. When the training was left to continue after 3 epochs, significant symptoms of overfitting began to appear.

6   Experimental Results and Evaluation

When fully trained, the classifier was evaluated on the *testing/evaluation set* in two different tasks: (i) **RAT**—the **relevance assessment task**, i.e. the input DOM node was classified into two classes only—a *relevant* content and *irrelevant* content; (ii) **FAT**—the **forum analysis task** where the classifier was used to identify components of the forum post (author, date, post text).

Fig. 2. Web pages of the discussion forums of the online student community **The Student Room**, https://www.thestudentroom.co.uk (left), and the vegetarian & vegan community **veggieboards**, https://www.veggieboards.com (right), as processed by the described relevance assessment technique (the darker background colour, the more relevant the content is).

In the **RAT** task, the original classes $\langle 1, 6 \rangle$ were grouped together and considered the *useful content* while the original class 0 was left intact as the *unwanted content*, or junk. Thus, the final decision was dichotomic: classes $\langle 1, 6 \rangle$ vs class 0.

The **FAT** task groups the original classes as follows: The **N** (name) class consists of the original classes 1 and 4. The **D** (date) class is 2 and 5. Finally, the **T** (text) class is 3 and 6. The junk class **J** equals to the original class 0. The classifier put the unknown input node into one out of four classes: **N**, **D**, **T**, and **J**.

The achieved results are expressed as the Macro-F_1 score on 90% level of significance. In order to quantitate the effect of entropy-like feature (ELF)

Table 3. Classifier performance as evaluated on two different tasks with 90% level of significance.

Task	Macro-F_1 score	
	Baseline	ELF injection
RAT	0.8090 ± 0.0028	**0.8173 ± 0.0046**
FAT	0.7402 ± 0.0059	**0.7530 ± 0.0061**

injection, the classifier performance was measured without (the **Baseline** column in Table 3) and with the ELF injected (the **ELF Injection** column in Table 3) into the penultimate layer of the DNN-based classifier.

It can be seen that there is an objectively established rise in the classification accuracy of about 1% in both tasks when compared to the baseline system without ELF injection.

6.1 Relevance Assessment Examples

Figure 2 shows graphically the effect of content relevance assessment. Two web pages of internet discussion forums are repainted according to the ELF value so that the darker the background colour is, the more relevant the node content is.

7 Conclusion

It was shown that injecting a meaningful global (or at least non-local) measure into a deep neural network-based classifier helps to achieve better performance. An *entropy-like feature* was used for this injection based upon considerations and observations that a measure of disorderedness is higher for the web page DOM nodes containing advertisements or similar kind of commercial content.

The proposed technique increased the accuracy (in terms of Macro-F_1 score) of the content relevance assessment by about 1%.

Acknowledgments. This work has been partly supported by ERDF Research and Development of Intelligent Components of Advanced Technologies for the Pilsen Metropolitan Area (InteCom) (no.: CZ.02.1.01/0.0/0.0/17 048/0007267); and by Grant No. SGS-2019-018 Processing of heterogeneous data and its specialized applications. Computational resources were supplied by the project "e-Infrastruktura CZ" (e-INFRA LM2018140) provided within the program Projects of Large Research, Development and Innovations Infrastructures.

References

1. Abadi, M., et al.: TensorFlow: large-scale machine learning on heterogeneous systems (2015). https://www.tensorflow.org/. Software available from tensorflow.org

2. Coalition for Better Ads: Ad experience: Ad density higher than 30%, February 2018. https://www.betterads.org/mobile-ad-density-higher-than-30/. Accessed 10 May 2019
3. Crescenzi, V., Mecca, G., Merialdo, P., et al.: Roadrunner: towards automatic data extraction from large web sites. In: VLDB, vol. 1, pp. 109–118 (2001)
4. Ekštein, K., Pavelka, T.: Entropy and entropy-based features in signal processing. In: Proceedings of International Ph.D. Workshop Systems and Control: Young Generation Viewpoint 2004 (2004)
5. Hochreiter, S., Schmidhuber, J.: Long short-term memory. Neural Comput. **9**(8), 1735–1780 (1997)
6. Huang, Z., Xu, W., Yu, K.: Bidirectional LSTM-CRF models for sequence tagging. arXiv preprint arXiv:1508.01991 (2015)
7. Kim, Y., Jernite, Y., Sontag, D., Rush, A.M.: Character-aware neural language models. In: Thirtieth AAAI Conference on Artificial Intelligence (2016)
8. Neculoiu, P., Versteegh, M., Rotaru, M.: Learning text similarity with Siamese recurrent networks. In: Proceedings of the 1st Workshop on Representation Learning for NLP, pp. 148–157 (2016)
9. Shannon, C.E.: A mathematical theory of communication. Bell Syst. Tech. J. **27**(3), 379–423 (1948)
10. Sido, J., Konopík, M.: Deep learning for text data on mobile devices. In: 2019 International Conference on Applied Electronics (AE), pp. 1–4. IEEE (2019)
11. Sido, J., Konopík, M., Pražák, O.: English dataset for automatic forum extraction. Computación y Sistemas **23**(3), 765–771 (2019)
12. Wang, S., Jiang, J.: Learning natural language inference with LSTM. In: Proceedings of the 2016 Conference of the North American Chapter of the Association for Computational Linguistics: Human Language Technologies, pp. 1442–1451 (2016)
13. Wikipedia Contributors: Document object model – Wikipedia, the free encyclopedia (2019). https://en.wikipedia.org/wiki/Document_Object_Model. Accessed 10 May 2019
14. Wikipedia Contributors: Web crawler – Wikipedia, the free encyclopedia (2019). https://en.wikipedia.org/wiki/Web_crawler. Accessed 10 May 2019
15. Xingjian, S., Chen, Z., Wang, H., Yeung, D.Y., Wong, W.K., Woo, W.C.: Convolutional LSTM network: a machine learning approach for precipitation nowcasting. In: Advances in Neural Information Processing Systems, pp. 802–810 (2015)

ABCD: Analogy-Based Controllable Data Augmentation

Shuo Yang$^{(\boxtimes)}$ and Yves Lepage

Graduate School of Information, Production and Systems, Waseda University,
Shinjuku, Japan
yangshuo@toki.waseda.jp, yves.lepage@waseda.jp

Abstract. We propose an analogy-based data augmentation approach
for sentiment and style transfer named Analogy-Based Controllable Data
Augmentation (ABCD). The object of data augmentation is to expand
the number of sentences based on a limited amount of available data. We
are given two unpaired corpora with different styles. In data augmentation, we retain the original text style while changing words to generate
new sentences. We first train a self-attention-based convolutional neural
network to compute the distribution of the contribution of each word
to style in a given sentence. We call the words with high style contribution style-characteristic words. By substituting content words and
style-characteristic words separately, we generate two new sentences. We
use an analogy between the original sentence and these two additional
sentences to generate another sentence. The results show that our proposed approach decrease perplexity by about 4 points and outperforms
baselines on three transfer datasets.

Keywords: Affective computing · Computing with words · Natural
language processing · Neural networks

1 Introduction

In recent years, artificial neural networks have been adopted in almost all fields of
natural language processing. Pre-trained language representation models [2,14]
showed that training on very large amounts of data can achieve state-of-the-
art in various tasks. However, for the style transfer and classification tasks,
unsupervised learning faces a serious problem with the lack of available data.
Due to the difficulty in manual data collection or data creation, automatic data
augmentation is a solution worthy of consideration. The method we introduce
in this paper mainly focuses on text pre-processing in the domain of natural
computing. It is essentially a carefully improved form of synonym-substitution-
based approach.

Data augmentation consists in creating an additional assortment of sentences
in the same domain as an input sentence while substituting the original words as
much as possible. These generated sentences are supposed to diverse downstream

C. C. Aranha et al. (Eds.): TPNC 2021, LNCS 13082, pp. 69–81, 2021.
https://doi.org/10.1007/978-3-030-90425-8_6

Fig. 1. An example of the substitutions in the word-level for the Yelp dataset. The red words are style-characteristic words, i.e., words with the source style attribute. The blue words are content words.

tasks, e.g., text classification, text style transfer and unsupervised machine translation. In light of related style transfer work is mainly about sentiment transfer [9, 10], we conduct experiments in Sect. 3 on three datasets of sentiment transfer.

Theoretically, a simple and effective method is to employ editing operations [23]. One of the traditional approaches is based on synonym substitution. However, this method may result in an insufficient number of new samples because of the limitation of the size of synonym sets. For instance, a positive sentence *"Good bar food."* can be replaced with *"Amazing bar food."* by synonym substitution, but it is almost impossible to generate a sentence like *"Wonderful fish sandwich."*.

For random substitutions and deletions at word-level [18], text affected by random noise lacks readability. For example, *"This place was very good."* (Positive) → *"This place was eggplant good."* (Meaningless). Another potential problem is that the style-characteristic word "good" can be deleted or replaced randomly (For the attribute of sentiment, "style-characteristic" is equal to "emotional"). In this situation, it is difficult to ensure that the original style remains in the generated text.

To avoid the above disadvantages, we employ two semantic analogy approaches and propose a novel data augmentation model based on reinforcement learning (RL). In practice, we first pre-train a classifier on labelled datasets. It is used to label style-characteristic words in a sentence and supervise the training of a style pointer network. For an input list of sentences $A = (a_1, a_2, ..., a_N)$, we train the style pointer by maximum likelihood estimation (MLE) to find out the word with the highest contribution in style in any unknown input sentence. Since the Transformer [16] performs well in varied domains, we employed the standard Transformer Encoder structure to build the style pointer network.

These style-characteristic words will be masked, the left word sequence will be regarded as a content representation of the input sentence.

Secondly, we follow the SeqGAN algorithm [22] to train another Transformer called a style painter network to generate a new style-characteristic word. In such a case, after we identify and mask a style-characteristic word, the trained style painter is used to fill in the mask. In conjunction, the style painter and the classifier will form a generative adversarial network (GAN). To make sure that the re-sampled word is suitable for the content representation, we train a 5-gram model to perform filtering of output sentences in each iteration. We name the final generated sentence list of the style painter as $B = (b_1, b_2, ..., b_N)$, where b_i is a new sentence.

As mentioned above, we argue that style-characteristic word substitution is not enough for data augmentation. Thus, we modify the content representation concurrently. We consider that pointwise mutual information (PMI) between words and styles can reflect the style contribution of words [13]. By computing the PMI between each word and each style independently, it is possible to find out style-independent words in the content representation. We assume that the substitution of a style-independent word will not result in a change in the style of the original sentence. Finally, we generate a second list of sentences $C = (c_1, c_2, ..., c_N)$ by using random substitution on these style-independent words.

For the input list of sentences A, the first list of sentence B and the second list of sentence C, we generate an additional list of sentence $D = (d_1, d_2, ..., d_N)$ by constructing the following analogical relationship based on [8]:

$$a_i : b_i :: c_i : d_i, \tag{1}$$

where a_i stands for the i-th sentence in the batch A (same for b_i and c_i), the solution d_i is supposed to consist of absolutely different style-characteristic words and content words from a_i. By regarding sentences as being word strings, the proposed approach generates analogies as the following one:

It	*It*	*This*	*This*
was truly	: *was truly* ::	*was truly*	: *was truly*
outstanding!	*amazing!*	*outstanding!*	*amazing!*

The above example shows positive sentence augmentation on the Yelp dataset, where the blue words are style-independent content words and the red words are style-characteristic words. Our contributions are as follows:

- We propose a novel approach for data augmentation which especially pays attention to text fluency.
- To the best of our knowledge, this paper is the first one that uses analogy in data augmentation for style transfer.
- The number of generated sentences of the proposed approach can be specified in advance. Since the generation based on the analogy can be repeated, the proposed method can substantially expand the number of samples. It means that the extent of the augmentation is controllable.

2   Methodology

We consider a list of sentences $X = (x^{(1)}, ..., x^{(N)})$ with the source style s_X and another list of sentences $Y = (y^{(1)}, ..., y^{(M)})$ where each $y^{(i)}$ is a sentence with the style s_Y. All of the used datasets in this paper are non-parallel, i.e., $N \neq M$. For each of these two sets, we generate a new batch of sentences to expand the original lists. For instance, $\hat{X} = \sum_{i=1}^{M} \sum_{j=1}^{k} \hat{x}_j^{(i)}$ is the set obtained at the k-th step of the augmentation process, where $\hat{x}_j^{(i)}$ is generated from $x^{(i)}$, i.e. they have the same style s_X. To simplify the presentation, the formalization in this paper focuses on the augmentation of X. In the following sections, the list of sentences A refers to the list of sentences X. In Sect. 3, the experiments are performed on both X and Y.

2.1   Overview

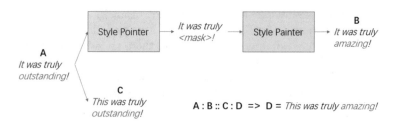

Fig. 2. The red tokens represent the style-independent content words, the blue tokens represent the style-characteristic words and the brown tokens represent the special symbols.

Text style is defined by style-characteristic words [9,21]. We propose that the substitution of style-characteristic words and content words should be considered independently.

For style-characteristic words, we train a style classifier on the two original corpora X and Y to calculate a distribution of the style contribution of words. The words with a high style contribution are labelled. With these labels, we train a style pointer network to identify style-characteristic words in an input sentence. We mask these words to get a content representation. To complete the content representation, we train a style painter network. It is used to fill in the masks. In this way, we rewrite all of the sentences in X with the same content but different style-characteristic words. We name the generated set as B.

For style-independent content words, we identify them by calculating the PMI between words and styles. By replacing content words in $x^{(i)}$ randomly, we generate a new sentence with different content words and the same style-characteristic words. We call the generated set C. To ensure the substitution does not make the sentences semantically contradictory, we pre-train a language model for supervision.

Finally, we generate another sentence with new style-characteristic words and new content words by analogy. This generated set D will be treated as the input set A for the next iteration. The overview of the proposed approach is shown in Fig. 2.

2.2 Style Pointer

We used the standard Transformer encoder architecture [16] to build the style pointer network. It replaces style-characteristic words in the input text with the special token $<MASK>$. The sentence with this special token will be filled into a complete sentence by the style painter network in the next step.

We first train a self-attention based convolutional neural network (CNN) as a classifier on the corpora. We assume that the style attribute of the sentence is affected by some words of the sentence. We try to delete each word in an input sentence one after another. The classifier is used to score the influence of the absence of a deleted word. It is based on the fact that if a deletion leads to a confidence drop in the logits of the classifier, then the word is stylistically relevant. Under this assumption, we delete each word from the sentence and calculate the accuracy for the rest of the sentence by using the classifier. We propose that a word with a higher style contribution leads to lower accuracy that the sentence without the word is inputted to a pre-trained classifier. The stylistic contribution word w_j is calculated by formula (2).

$$\eta(w_j) = P(x; \phi) - P(x'; \phi), \tag{2}$$

where ϕ is the parameters of the classifier, x' is composed of x delete the word w_j.

We label the style-characteristic words in each sentence from the source corpus. We train the style pointer on the labelled data by using MLE. The final loss function is (3). The trained style pointer is used to make word judgments on future inputs. The algorithm for training the style pointer network is shown in 1.

$$L(\theta) = -\frac{1}{M} \sum_{i=1}^{M} \log P(j|x; \theta), \tag{3}$$

where M is the number of samples in the dataset X, θ is the style pointer network, j is the labelled style-characteristic word in the sentence x.

2.3 Style Painter

To ensure that the style of a sentence is preserved, we build a standard Transformer as the style painter network to fill in the masks with re-sampled style-characteristic words. A similar algorithm of SeqGAN [22] is used for training. Different from it, we do not need a pseudo-parallel corpus for supervision and pre-training. In our approach, the loss of the style painter network is totally from the style classifier.

Algorithm 1. Learning Method for Training the Style Pointer Network θ

1: Pre-train a binary style classifier ϕ.
2: **for** each iteration **do**
3:     Sample a sentence x with the source style s_X a from X
4:     **for** each token w_j in x **do**
5:         Score the style contribution of w_j by using (2)
6:     **end for**
7:     **if** $\eta(w_j) \leq$ Threshold **then**
8:         Label the position j
9:     **end if**
10: **end for**
11: **for** each training iteration **do**
12:     Sample a sentence x with the source style s_X and the corresponding label j
13:     Update θ by following (3)
14: **end for**

We regard the classifier as a discriminator and regard the style painter as a generator. Hence, a GAN is formed with the style painter and the classifier. Due to the characteristics of natural language, backpropagation cannot be employed directly [21]. Therefore, we use the policy gradient [19] to updating the parameters of the style painter network.

$$L(\omega) = -\mathbb{E}[R(x') \cdot \log P(w_j|x';\omega)], \tag{4}$$

where ω is the parameters of the style painter network and $R(x')$ is the reward value of sentence x' calculated by the style classifier. The reward function is given in (5).

$$R(x') = P(s_Y|x';\phi), \tag{5}$$

where ϕ is the style classifier and x' is a generated sentence.

Another advantage is that the generator trained by reinforcement learning can increase the probability of the style-characteristic words being sampled. To expand the selection range of these words, Nucleus Sampling [4] is used to fill in the mask. An example of attribute substitution in the word level is shown in Fig. 1.

On the other hand, the policy gradient is employed to increase the likelihood of appropriate actions sampled. This is Algorithm 2.

2.4   Pointwise Mutual Information (PMI)

We compute the PMI between words and different styles to point to style-independent content words. It is based on the assumption that if a word has similar values of PMI for different styles, this word is probably style-independent. An example of content substitution in the word level is shown in Fig. 1.

Algorithm 2. Learning Method for Training the Style Painter Network

---
1: Pre-train a binary classifier ϕ
2: Training an n-gram language model on two corpora
3: **for** each iteration **do**
4:    Sample a sentence x with style s_X a from X
5:    Generate a sentence $\hat{x}$ by using the style painter network ω
6:    **for** $j = (1, ..., N)$ **do**
7:       Extract a word sequence $(\hat{w}_1, \hat{w}_2, ..., \hat{w}_j)$
8:       Complete this word sequence to a sentence x' by using the style painter network ω
9:       Compute a reward $R(x')$ of x' by following Formula (5)
10:    **end for**
11:    Update ω by following Formula (4)
12: **end for**

For the word w_j, we propose the formula 6 by calculating the arithmetic mean of the PMI between a word w_j and two styles.

$$\text{PMI}(w_j; s_X, s_Y) = \log \frac{\sqrt{p(w_j|X) \cdot p(w_j|Y)}}{p(w_j|X+Y)}, \tag{6}$$

where $W(X_i)$ is the number for words in the dataset X_i and $p(w|X) = C(w)/W(x)$, where $C(\cdot)$ is the number of times w appears in X.

The style independent words should have smaller I values than the other words. We select the bottom k_i words as the replacement scheme of content words for the input sentences.

2.5 Analogy

Before we generate the sentence d_i, an issue we should consider first is whether the substituted words in b_i and c_i are conflict. If we just consider the formal analogy between character strings, the output sentences may not be semantically fluent, which can lead to an increase in perplexity, e.g., *'The icing is amazing.'* : *'The icing is delicious.'* :: *'The service is amazing.'* : *'The service is delicious.'*.

Formal analogies do not care about meaning. To ensure that there are no semantic mistakes in the generated sentences, we solve semantic analogies. The two papers most similar to our work are [7] and [17].

For the first approach, we use semantico-formal analogy [7]. It combines formal analogies between strings and semantic analogies between words. An assumption is that the Euclidean distance between words in the embedding space represents the semantic distance between words. Traces are explored between the inputs to reach resolutions by formal analogies [6].

For the second approach, we use a semantic analogy between sentences. Motivation is that, if sentences a_i, b_i and c_i can be used to solve the resolution d_i, then it should have a set of mapping functions $\mathcal{F}$ for these sentences embedding vectors in the latent space. With the analogy relation for a finite number of samples $a_i : b_i :: c_i : d_i$, we can always find at least one common mapping function $F \in \mathcal{F}$ to follow the formula (7).

$$F(\vec{a_i}) = \vec{b_i} \rightarrow F(\vec{c_i}) = \vec{d_i}, \tag{7}$$

where $\vec{x_i}$ stands for the sentence vector representation for x_i.

In the previous work [17], the sentence embedding vector $\vec{d_i}$ is computed on $\vec{d_i} = \vec{b_i} - \vec{a_i} + \vec{c_i}$. This is arithmetic analogy between vectors.

In this paper, we do not introduce any additional rule to calculate the function F. We propose a simple approach that trains a seq2seq model to learn the mapping function directly from $\vec{a_i}$ and $\vec{b_i}$ by formula (7). The learned F is used to generated $\vec{d_i}$ from $\vec{c_i}$.

3   Experiments

3.1   Dataset

For all used datasets, we filtered out sentences with hapax legomena, i.e., a word with only one instance of use (Table 1).

Table 1. Statistic of the used datasets.

Dataset	Yelp		IMDb		SST-2	
Category	Positive	Negative	Positive	Negative	Positive	Negative
Number of samples	266041	177218	178869	187597	256	244
Average length	8.43	9.55	18.90	18.31	17.70	17.89

Yelp. The Yelp dataset contains remarks of consuming spots from Yelp users. It is a positive and negative reviews collection.

IMDb. The IMDb Movie Review Dataset dataset contains positive and negative reviews of movies. We use the cleaned version provided by [1], which is reported created on the basis of previous work [12].

SST-2. The Stanford Sentiment Treebank v2 [15] is used to perform the augmentation of a small amount of data in this paper. We used the cleaned version [18] of the SST-2. It contains 500 sentences.

3.2   Evaluation Metric

Accuracy. The value of accuracy indicates the style preservation for the generated text. We pre-train a binary classifier to measure generation. A higher accuracy means that the expanded sentence has the same style as the original input sentence.

$$\text{Accuracy} = \frac{1}{M} \sum_{i}^{M} P(s_X | x^{(i)}; \phi) \tag{8}$$

Perplexity. We build a 5-gram language model on the source dataset using KenLM [3]. This language model is used to compute text fluency by calculating the perplexity of the generated sentences. A lower perplexity indicates a higher text fluency.

3.3 Details

The used pointer is a Transformer that consists of 4 layers of basic units. Each layer has 8 attention heads and 512 dimensions. The used painter is a Transformer with 6 layers in the encoder and 6 layers in the decoder. Each of them has the same parameters as the layers in the pointer module. We train a classifier with three different convolutional kernel sizes of 2, 3, 4 for evaluation. For the training step, we employ the Adam algorithm [5] to learn models with a learning rate of 0.0001.

4 Results

Table 2. Automatic evaluation results of the three times augmentation. The "Self Input" is the test result of the sentences in the input dataset.

Dataset	Yelp		SST-2		IMDb	
Model	Accuracy	Perplexity	Accuracy	Perplexity	Accuracy	Perplexity
Self input	N/A	8.75	N/A	15.70	N/A	20.23
SynoReplace [23]	88.17	53.98	98.94	20.56	92.78	96.93
EDA [18]	**96.93**	75.56	**99.60**	33.21	92.7	99.78
DataNoising [20]	90.64	87.96	90.63	211.26	**92.88**	115.39
BackTrans [24]	94.71	86.93	89.84	116.16	91.01	142.25
ABCD+Vec2Sequence	94.79	55.60	84.34	29.40	89.61	169.45
ABCD+Semantico-analogy	95.63	**51.93**	98.17	**20.34**	91.59	**92.16**

The results show that our model outperforms the baselines on text fluency on the 3 datasets. However, we notice that the accuracy is not very high. Related work with style transfer on these datasets [1, 11] reported that accuracy may not be an effective metric because even the human evaluation results do not achieve high accuracy.

For the Yelp dataset, both of our proposed approaches achieve relatively high accuracy and low perplexity. The accuracy of the Semantical-analogy model is 95.63, similar to the EDA model. The perplexity of 51.93 is lower than all the used baselines, similar to the synonym-replaced approach. Nevertheless, the analogy created by training a generation model directly is not as good as the one created by using the semantico-analogy model. We suppose that two reasons explain this phenomenon. The first one is that the functions to perform analogies between different sentences pairs may be different in the latent space. The

Table 3. Examples of generated sentences. We test on all the baselines from Table 2 to perform 3 times data augmentation. The blue words indicate inappropriate words. The red words represent appropriate word substitutions. The brown sentences indicate no word is changed.

Dataset	Yelp
Input	they did a really good job and i enjoyed my experience there
SynoReplace [23]	they did a trully good job and i enjoyed my experience there
EDA [18]	they did a there good job and i enjoyed my experience really
DataNoising [20]	they did a really good job and i the my experience there
BackTrans [24]	they did a good job and I enjoyed the experience there
ABCD+Vec2Sequence	they did a really professional job and i enjoyed my experience there
ABCD+Semantical-analoy	they did a really fun job and i enjoyed my visit there
Dataset	IMDb
Input	i can not recommend it highly enough
SynoReplace [23]	i can not recommend it highly adequate
EDA [18]	i can not it recommend highly enough
DataNoising [20]	i can not recommend it of enough
BackTrans [24]	i cannot recommend it highly
ABCD+Vec2Sequence	i can not just it highly enough
ABCD+Semantical-analoy	kids can not expect it highly enough
Dataset	SST-2
Input	A true delight
SynoReplace [23]	a truthful delight
EDA [18]	a true up true delight
DataNoising [20]	a true imagine
BackTrans [24]	a true delight
ABCD+Vec2Sequence	a true delight
ABCD+Semantical-analoy	the true movie

objective function may be too complex to be learned by the generation model when the number of parameters of the used Transformer is insufficient. This happens when the generation model gets caught in a local optimal solution, resulting in poor performance. Another one is that the generation of one sample is influenced by other samples. For the training process, if the quality of part of the input data is not good enough, a generator will be affected by these noises.

In contrast, for the semantico-analogy, a neural-independent approach has no training process, so the generation of input sentences is virtually independent.

For the SST-2 dataset, problems with the neural networks cause the semantico-analogy to perform poorly. When the training data is insufficient, it is difficult for the generator to learn a suitable map function. Also, the generated text faces the problem of text degeneration. Due to the iterative relation between inputs and outputs in our model, this error may gradually accumulate as text is continuously generated. However, as expected, the results of the Vec2Sequence model is comparable with an accuracy of 84.34 and a perplexity of 29.40.

For the IMDb dataset, all the models show a similar accuracy and high perplexity. We presume that the IMDb dataset has longer sentences on average and a smaller number of samples than the Yelp dataset. This results in the pre-trained classifier and LM being relatively unreliable. The semantical-analogy model still shows a low perplexity with a value of 92.16.

We sample sentences randomly from the datasets to demonstrate the very high readability of our generated text. This is Table 3. Our model outperforms baselines in fluency. We show two sentences in brown colour which means no word is changed from the inputs. We do not evaluate these sentences in our experiments because they cannot represent a successful data augmentation.

5 Conclusions

In this paper, we proposed a novel analogy-based approach to data augmentation. Compared with related works, the proposed model performs well for the fluency of the generated text. Besides, we proposed methods for searching and substituting style-characteristic words and content words separately, which allows our model to be compatible with sentence analogy. Since the analogy-based method is iterative, the proposed model has a strong generative capability. Furthermore, the proposed model is not limited to the data augmentation tasks. Substitution rules for style can be learned by the style pointer network. Such rules can be used directly for other controllable language generation tasks. In future work, we will examine the use of n-grams instead of individual words to inspect the possibility of rewriting sentences more flexibly.

References

1. Dai, N., Liang, J., Qiu, X., Huang, X.: Style transformer: unpaired text style transfer without disentangled latent representation. In: Proceedings of the 57th Annual Meeting of the Association for Computational Linguistics, pp. 5997–6007. Association for Computational Linguistics, Florence, Italy, July 2019. https://doi.org/10.18653/v1/P19-1601
2. Devlin, J., Chang, M.W., Lee, K., Toutanova, K.: BERT: pre-training of deep bidirectional transformers for language understanding. In: Proceedings of the 2019 Conference of the North American Chapter of the Association for Computational Linguistics: Human Language Technologies, pp. 4171–4186. Association for Computational Linguistics, Minneapolis, June 2019. https://doi.org/10.18653/v1/N19-1423

3. Heafield, K.: KenLM: faster and smaller language model queries. In: Proceedings of the Sixth Workshop on Statistical Machine Translation, pp. 187–197. Association for Computational Linguistics, Edinburgh, July 2011

4. Holtzman, A., Buys, J., Du, L., Forbes, M., Choi, Y.: The curious case of neural text degeneration. In: International Conference on Learning Representations (2020)

5. Kingma, P.D., Ba, L.J.: Adam: a method for stochastic optimization. In: International Conference on Learning Representations (2015)

6. Lepage, Y.: Solving analogies on words: an algorithm. In: 36th Annual Meeting of the Association for Computational Linguistics and 17th International Conference on Computational Linguistics, vol. 1, pp. 728–734. Association for Computational Linguistics, Montreal, August 1998. https://doi.org/10.3115/980845.980967

7. Lepage, Y.: Semantico-formal resolution of analogies between sentences. Proceedings of LTC, pp. 57–61 (2019)

8. Lepage, Y., Ando, S.I.: Saussurian analogy: a theoretical account and its application. In: COLING 1996 Volume 2: The 16th International Conference on Computational Linguistics (1996)

9. Li, J., Jia, R., He, H., Liang, P.: Delete, retrieve, generate: a simple approach to sentiment and style transfer. In: Proceedings of the 2018 Conference of the North American Chapter of the Association for Computational Linguistics: Human Language Technologies, Volume 1 (Long Papers), pp. 1865–1874. Association for Computational Linguistics, New Orleans, June 2018. https://doi.org/10.18653/v1/N18-1169

10. Li, X., Chen, G., Lin, C., Li, R.: DGST: a dual-generator network for text style transfer. In: Proceedings of the 2020 Conference on Empirical Methods in Natural Language Processing (EMNLP). Association for Computational Linguistics, November 2020

11. Luo, F., et al.: A dual reinforcement learning framework for unsupervised text style transfer. In: Proceedings of the Twenty-Eighth International Joint Conference on Artificial Intelligence, IJCAI-19, pp. 5116–5122. International Joint Conferences on Artificial Intelligence Organization, July 2019. https://doi.org/10.24963/ijcai.2019/711

12. Maas, A.L., Daly, R.E., Pham, P.T., Huang, D., Ng, A.Y., Potts, C.: Learning word vectors for sentiment analysis. In: Proceedings of the 49th Annual Meeting of the Association for Computational Linguistics: Human Language Technologies, pp. 142–150. Association for Computational Linguistics, Portland, June 2011

13. Pavlick, E., Nenkova, A.: Inducing lexical style properties for paraphrase and genre differentiation. In: Proceedings of the 2015 Conference of the North American Chapter of the Association for Computational Linguistics: Human Language Technologies, pp. 218–224. Association for Computational Linguistics, Denver, May-June 2015. https://doi.org/10.3115/v1/N15-1023

14. Radford, A., Wu, J., Child, R., Luan, D., Amodei, D., Sutskever, I.: Language models are unsupervised multitask learners. OpenAI Blog 1(8), 9 (2019)

15. Socher, R., et al.: Recursive deep models for semantic compositionality over a sentiment treebank. In: Proceedings of the 2013 Conference on Empirical Methods in Natural Language Processing, pp. 1631–1642. Association for Computational Linguistics, Seattle, October 2013

16. Vaswani, A., et al.: Attention is all you need. In: NIPS (2017)

17. Wang, L., Lepage, Y.: Vector-to-sequence models for sentence analogies. In: 2020 International Conference on Advanced Computer Science and Information Systems (ICACSIS), pp. 441–446 (2020). https://doi.org/10.1109/ICACSIS51025.2020.9263191

18. Wei, J., Zou, K.: EDA: easy data augmentation techniques for boosting performance on text classification tasks. In: Proceedings of the 2019 Conference on Empirical Methods in Natural Language Processing and the 9th International Joint Conference on Natural Language Processing (EMNLP-IJCNLP), pp. 6382–6388. Association for Computational Linguistics, Hong Kong, November 2019. https://doi.org/10.18653/v1/D19-1670
19. Williams, R.J.: Simple statistical gradient-following algorithms for connectionist reinforcement learning. Mach. Learn. **8**(3–4), 229–256 (1992)
20. Xie, Z., et al.: Data noising as smoothing in neural network language models. In: 5th International Conference on Learning Representations, ICLR 2017, January 2019
21. Xu, J., et al.: Unpaired sentiment-to-sentiment translation: a cycled reinforcement learning approach. In: Proceedings of the 56th Annual Meeting of the Association for Computational Linguistics (Volume 1: Long Papers), pp. 979–988. Association for Computational Linguistics, Melbourne, July 2018. https://doi.org/10.18653/v1/P18-1090
22. Yu, L., Zhang, W., Wang, J., Yu, Y.: SeqGAN: sequence generative adversarial nets with policy gradient. In: AAAI (2017)
23. Zhang, X., Zhao, J.J., LeCun, Y.: Character-level convolutional networks for text classification. In: NIPS (2015)
24. Zhang, Y., Ge, T., Sun, X.: Parallel data augmentation for formality style transfer. In: Proceedings of the 58th Annual Meeting of the Association for Computational Linguistics, pp. 3221–3228. Association for Computational Linguistics, July 2020. https://doi.org/10.18653/v1/2020.acl-main.294

Evolutionary and Swarm Algorithms

MOEA/D with Adaptative Number of Weight Vectors

Yuri Lavinas[✉], Abe Mitsu Teru, Yuta Kobayashi, and Claus Aranha

University of Tsukuba, Tsukuba, Japan
{lavinas.yuri.xp,abe.mitsuteru.xw,
kobayashi.yuta.xu}@alumni.tsukuba.ac.jp,
caranha@cs.tsukuba.ac.jp

Abstract. The Multi-Objective Evolutionary Algorithm based on Decomposition (MOEA/D) is a popular algorithm for solving Multi-Objective Problems (MOPs). The main component of MOEA/D is to decompose a MOP into easier sub-problems using a set of weight vectors. The choice of the number of weight vectors significantly impacts the performance of MOEA/D. However, the right choice for this number varies, given different MOPs and search stages. We adaptively change the number of vectors by removing unnecessary vectors and adding new ones in empty areas of the objective space. Our MOEA/D variant uses the Consolidation Ratio to decide when to change the number of vectors and to decide where to add or remove these weighted vectors. We investigate the effects of this adaptive MOEA/D against MOEA/D with a poorly chosen set of vectors, a MOEA/D with fine-tuned vectors and MOEA/D with Adaptive Weight Adjustment on two commonly used benchmark functions. We analyse the algorithms in terms of hypervolume, IGD and entropy performance. Our results show that the proposed method is equivalent to MOEA/D with fine-tuned vectors and superior to MOEA/D with poorly defined vectors. Thus, our adaptive mechanism mitigates problems related to the choice of the number of weight vectors in MOEA/D, increasing the final performance of MOEA/D by filling empty areas of the objective space and avoiding premature stagnation of the search progress.

Keywords: MOEA/D · Auto adaptation · Multi objective optimisation

1 Introduction

Multi-objective Optimisation Problems (MOPs) have multiple objectives with trade-off relationships making it hard to find a single solution that provides a good balance too all objectives. Thus there is a need to find a set of trade-off solutions, called Pareto Front (PF) solutions. A critical characteristic of these solutions is to cover all regions of the optimal PF, without any sparsity regions[1].

---

[1] In the case of discontinuous PF case, there is no need to cover the discontinuity area.

© Springer Nature Switzerland AG 2021
C. C. Aranha et al. (Eds.): TPNC 2021, LNCS 13082, pp. 85–96, 2021.
https://doi.org/10.1007/978-3-030-90425-8_7

These empty areas of the PF indicate that different possible trade-offs are still to be found.

One of the most common algorithms for finding good sets of solutions for MOPs is the Multi-Objective Evolutionary Algorithm Based on Decomposition (MOEA/D) [15]. The most crucial feature of MOEA/D is that this algorithm decomposes the MOP into several single-objective problems. This decomposition of the MOP depends on a set of *weight vectors*, where each vector corresponds to a different region of the PF. The choice of weight vectors is essential on MOEA/D and the appropriate value generally is not known, and are closely related to the population size, influencing its dynamic during the search progress. This dynamic also depends on the difficulty of the problem, the presence of multiple local optima, the shape of the PF and other features [1,5]. Thus, using a low number of vectors may lead to search stagnation, while a high number may waste computational resources.

A growing body of literature recognises the need to define the appropriate set of weight vectors in MOEA/D [4,7,10–12]. One major issue in these works is that they focus on adjusting the position of weight vectors in terms of the objective space, paying little attention to defining the number of weight vectors. In summary, much uncertainty still exists about the relationship between redefining the weight vectors adaptively in MOEA/D and the coverage of the empty spaces of the PF region while avoiding premature convergence.

Here, we focus on automatically adapting the number of weight vectors in MOEA/D, adding or deleting vectors based on the progress of the search. Our proposed adaptation method has two main components: (1) to identify the timing to add and remove weight vectors, and (2) to decide which vectors to add or remove. To identify the timing to change vectors, we use the *Consolidation Ratio*, which was initially proposed as an online stopping criterion [6]. To decide which vectors to add or remove, we use two strategies, one based on uniform sampled values and the other based on the adaptive weight adjustment (AWA) [12]. Moreover, our method dependency on the initial set of weight vectors is small and given its adaptive nature requiring little fine-tuning of the number of vectors.

The proposed method was tested on the DTLZ and ZDT benchmark and compared with (1) MOEA/D with different weight vector settings and (2) MOEA/D-AWA [12], a method that adjusts the positions of the weight vectors during the search. This study provides new insights into the relationship between the choice of the number of weight vectors in MOEA/D and the increments of performance of MOEA/D by filling empty areas of the objective space while avoiding premature stagnation of the search progress.

2   MOEA/D and Weight Vectors

The MOEA/D algorithm is characterized by the decomposition of the MOP into many sub-problems. A sub-problem is characterized by a weight vector λ highly influencing the performance of MOEA/D. When adding new weight vectors, it is necessary to decide where a new set of vectors should be positioned.

Several works study the problem of changing the positions of weight vectors in MOEA/D. For more detailed information, refer to this Survey [11]. One of the most popular methods for guiding the weight adaptation strategy is the Adaptive Weight Adjustment (AWA) [4,7,10,12]. A major advantage of AWA is that it changes the position of the vectors according to the feature of the MOPs and that is why we use AWA as one of the base mechanisms in this work.

The AWA method keeps an external archive and re-position the vectors to the sparsest regions of this archive. This positioning is based on the Sparsity Level (SL) for each solution in the archive:

$$SL(ind_j, pop) = \prod_{i=1}^{m} L_2^{NN_i^j} \qquad (1)$$

where $L_2^{NN_i^j}$ is the Euclidean distance between the j-th solution and its i-th nearest neighbour. The new vector, $\lambda^{\mathbf{sp}}$, is generated using the individual with the highest SL as based for calculation:

$$\lambda^{\mathbf{sp}} = \left(\frac{\frac{1}{\mathbf{f}_1^{\mathbf{sp}} - \mathbf{z}_1^* + \epsilon}}{\sum_{k=1}^{m} \frac{1}{\mathbf{f}_k^{\mathbf{sp}} - \mathbf{z}_k^* + \epsilon}}, ..., \frac{\frac{1}{\mathbf{f}_m^{\mathbf{sp}} - \mathbf{z}_m^* + \epsilon}}{\sum_{k=1}^{m} \frac{1}{\mathbf{f}_k^{\mathbf{sp}} - \mathbf{z}_k^* + \epsilon}} \right) \qquad (2)$$

Where f_k^{sp} is the objective function value of the individual with the biggest SL, z_k^* is the reference point of this same individual, and ϵ is a small number. This paper uses a modified version that considers the Unbounded External Archive (UEA) instead of the traditional external archive. For more information see Algorithm 2 and the original paper [12].

The proposed method uses the stagnation state of the search as criteria to decide when to add or remove weight vectors. Several studies study stagnation metrics in MOEAs as stopping criterion [6,14]. Here, we choose the CR indicator [6], since it requires no problem-dependent parameters.

The CR indicator uses the non-dominated archive A_i of the population at generation i, the non-dominated archive at generation $i - \Delta$, and the set S of solutions from $A_{i-\Delta}$ that are not dominated by A_i, and is calculated as

$$CR = \frac{|S|}{|A_i|},$$
$$S = \{a_{i-\Delta} : a_{i-\Delta} \not\prec a_i\}, \qquad (3)$$
$$a_i \in A_i, a_{i-\Delta} \in A_{i-\Delta}.$$

Using this indicator, we can compute the utility function U_i, and the average generation utility U_i^* (Eqs. 6 and 4). The decision to add or remove weights is made when U_i^* exceeds a threshold U_{thresh}:

$$U_i^* = \frac{U_i + U_{(i-\Delta)}}{2} \qquad (4)$$

$$U_{thresh} = \frac{U_{init}}{F} \qquad (5)$$

The F value were chosen based on the original CR paper [6]. U_{init} is calculated once, after the CR exceeds 0.5 for the first time, at generation G. CR is calculated as:

$$U_i = \frac{CR_i - CR_{(i-\Delta)}}{\Delta} \tag{6}$$

Lastly, U_{init} is only calculated once, after $CR > 0.5$ happens for the first time, at generation G. U_{init} is defined as:

$$U_{init} = \frac{CR}{G} \tag{7}$$

3   MOEA/D with Adaptive Weight Vectors

We propose a method to enhance MOEA/D by automatically adding or removing weight vectors as the search progresses, named MOEA/D-AV (**MOEA/D** with **A**daptive weight **V**ectors). The outline of this method is described in Algorithm 1. The code for the method and experiments in this study is available at a GitHub repository[2].

Now we explain the most relevant details of MOEA/D-AV. At every generation, this algorithm calculates the CR value. When the value of CR is larger than 0.5, the method moves on to calculate the threshold value U_{thresh}. Then at every generation, MOEA/D-AV calculates the average generation utility (Eq. 4). If it exceeds the value of U_{thresh}, then the algorithm adds new weight vectors. Otherwise, it deletes weight vectors. The number of vectors added or removed at each update is a fraction of the total number of weight vectors.

When adding new vectors, MOEA/D-AV has a choice of adding vectors using two methods: the first is based on AWA (Algorithm 2) and the second is based on uniform sampled values. This second method samples values from a uniform distribution and then generates new weight vectors (Algorithm 3). This second method is used because determining the position of new vectors only with AWA leads to early stagnation of the search.

The choice of which method to use is controlled by the p probability value, which changes as the search progresses. This probability is calculated as $p = \frac{n\_fe}{n\_eval}$. Where n_fe is the current number of function evaluations, and n_eval is the total evaluation budget. This equation causes MOEA/D-AV to add weight vectors generated from values sampled from a uniform distribution at the beginning of the search. Then, this MOEA/D variant is more likely to create new weight vectors using the AWA-based method at later states.

Besides adding weight vectors, MOEA/D-AV also deletes weight vectors, with the goal of avoiding wasting computational resources when there are too many weight vectors in use. Currently, the weight vectors are selected to be deleted randomly, excluding those weight vectors associated with the axis of each objective. Algorithm 4 describes the method in detail.

---

[2] https://github.com/YUYUTA/MOEADpy

Algorithm 1. MOEA/D-AV

Input: vector adaptation **ratio**, initial weight vectors **W**, MOEA/D variables (set of solutions, neighbourhood solution matrix,...)
Output: Unbounded External Archive **UEA**
 1: **UEA** $\leftarrow \emptyset$
 2: Initialize and evaluate population $\mathbf{X}^{(0)}$
 3: Update **UEA**
 4: set $CR_{Gen} \leftarrow$ NULL, $U_{thresh} \leftarrow$ NULL
 5: **while** termination criterion is not meet **do**
 6:     Generate new population $\mathbf{X}'^{(\mathbf{Gen})}$ and evaluate this new population
 7:     Update **UEA** & Select next population
 8:     **if** CR_{Gen} is NULL **then**
 9:         Calculate CR_{Gen}.
10:     **end if**
11:     **if** $CR_{Gen} > 0.5$ **then**
12:         **if** U_{thresh} is NULL **then**
13:             Calculate U_{thresh}
14:         **end if**
15:         **if** $U^*_{Gen} > U_{thresh}$ **then**
16:             $p = \frac{n\_fe}{n\_eval}$
17:             **nav = ratio * size(W)**
18:             **if** $p > random$ **then**
19:                 Add vectors using the Unbounded version of AWA (Algorithm 2)
20:             **else**
21:                 Add vectors using the Uniform selection method (Algorithm 3)
22:             **end if**
23:         **end if**
24:     **else**
25:         Delete vectors (Algorithm 4)
26:     **end if**
27: **end while**

Algorithm 2. Adding Vectors - method 1 (adds vectors based on the sparsity of the UEA)

Input: Unbounded External Archive **UEA**, current population **pop**, number of to add vectors **nav**
Output: Updated population **pop**
 1: set count $= 0$
 2: calculate SL of individual in **UEA** using Eq. 1
 3: **while** $count <$ **nav do**
 4:     set $\mathbf{ind^{sP}} = (\mathbf{x^{sP}}, \mathbf{FV^{sP}})$ which has largest SL
 5:     generate new vector λ^{sP} using Eq. 2
 6:     add $(\mathbf{ind^{sP}}, \lambda^{sP})$ to **pop**
 7:     count $=$ count$+1$
 8: **end while**
 9: **return pop**

Algorithm 3. Adding Vectors - method 2 (adds vectors based on values sampled from a uniform distribution)

Input: Unbounded External Archive **UEA**, current population **pop**, number of vectors to add **nav**
Output: Updated population **pop**
1: set count = 0
2: **while** *count* < **nav do**
3:     generate new vector $\lambda^{\mathbf{rand}}$ using values sampled from a uniform distribution
4:     set $\mathbf{ind^{rand}} = (\mathbf{x^{rand}}, \mathbf{FV^{rand}})$ best solution for $\lambda^{\mathbf{rand}}$ in **UEA**
5:     add $(\mathbf{ind^{rand}}, \lambda^{\mathbf{rand}})$ to **pop**
6:     count = count+1
7: **end while**
8: **return pop**

Algorithm 4. Delete Vectors

Input: Current population **pop**, number of vectors to delete **nav**
Output: Updated population **pop**
1: set count = 0
2: **while** *count* < **nav do**
3:     let $W_{notEdge}$ be W without the edge vectors
4:     randomly select **argument arg** from $W_{notEdge}$
5:     delete $(\mathbf{ind^{arg}}, \lambda^{\mathbf{arg}})$ from **pop**
6:     count = count+1
7: **end while**
8: **return pop**

4   Weight Vectors Experiment

We investigate the relationship between adapting the number of weight vectors, represented by MOEA/D-AV, and increments of performance in practice. Here, we compare the UEA of MOEA/D-AV, MOEA/D-DE, and the MOEA/D-AWA (which adjust the values of the weight vectors, but not their numbers). The three methods are compared with different numbers of initial weight vectors to analyze how they interact given this different settings.

The algorithms were compared on the Hypervolume (HV)[3], Inverted Generational Distance (IGD) and Entropy metrics. The first two comparison methods are common in the MOP literature. This Entropy metric [3] is beneficial to evaluate the sparseness of the PF and it is uses to measure how well a method deals with empty areas of the PF, the main motivation of this work. For the comparisons, we use the DTLZ benchmark set (3-objective, 10 dimensions) [2], and the ZDT set (2-objective, 30 dimensions) [16]. For a fair comparison, the algorithms are evaluated based on their Unbounded External Archive (UEA) [13].

---

[3] For the HV calculation, we use the reference point set to $(1 + 1/H, 1 + 1/H)$ for two objective problems and $(1 + 1/H, 1 + 1/H, 1 + 1/H)$ for three objective problems.

The general MOEA/D-DE parameters were are used here as they were introduced in [9]. On the other hand, the Generation gap and user-controlled fraction
are the same as from the CR paper [6]. Also, specific parameters of the MOEA/D-
AWA can be found at [12]. Finally, our MOEA/D-AV adds one new parameter,
the vector adaptation ratio. The number of weight vectors for the 3-objective
DTLZ benchmark set was selected to be {10, 21, 45, 105, 190, 496 and 990}.
The number of weight vectors for the 2-objective ZDT set was set to {10, 20,
50, 100, 200, 500 and 1000}[4]. We set the number of evaluations to 75000 and
the number of trials to 21. The difference in performance for each experimental
condition, across all benchmark sets, was evaluated using a two-sided Wilcoxon
signed-rank test paired by benchmark, with $\alpha = 0.05$.

5   Results

This section compares MOEA/D-AV against the traditional MOEA/D-DE and
MOEA/D-AWA (AWA for simplicity) with different numbers of weight vectors.
We recall that the algorithms are evaluated based on their Unbounded External
Archive (UEA) and not their final population [13], for fair comparisons. The
results of the statistical tests (Wilcoxon signed-rank test paired by benchmark,
with $\alpha = 0.05$) are shown on Table 1. In this work, we use the symbols "=", "+"
and "−" to indicate the results of the statistical test. The symbol "=" indicates
no statistically significant difference between the methods, while "+" is used
to indicate a significant difference in favour of MOEA/D-AV and "−" indicates
difference against MOEA/D-AV.

Table 1, left side, shows the mean and standard deviation of the *best setting
for each algorithm in terms of the number of weight vectors*, based on hypervolume values. The best method for each MOP is highlighted in shades of gray.
Looking at this Table, we can see that the best results are similar in terms of
HV and IGD. This result suggests that there is no apparent difference between
these methods. In terms of Entropy values, the proposed method performs a
little worse than the other methods, especially for DTLZ1-4. On the other hand,
Table 1, right side, shows the mean and standard deviation of the *worst setting
for each algorithm in terms of the number of weight vectors*. The results are
shown by this side of the Table 1 indicating that the MOEA/D-AV performs
better than the other two MOEA/D variants in all metrics. We understand that
the reason for this is that our method can compensate for initial bad choices of
the number of vectors and achieve competitive results. Table 2 shows the number
of vectors for best and worst settings scenarios in terms of HV.

It is interesting to note that using extremes values for the number of weight
vectors, such as 10 or 990, lead to bad HV performance, as can be seen in
Table 2. In the case of a higher number of weight vectors, a possible cause for
this low performance is due to the number of vectors being too large for the
algorithms to efficiently progress with the search progress. In the case of the

---

[4] We initialize the weight vectors using the Simplex-lattice Design (SLD) method,
causing the number to slightly change between MOPs with 2 and 3 objectives.

Table 1. Mean and standard deviation, in parenthesis, for all algorithms.

	Best scenario			Worse scenario		
HV	MOEA/D	AWA	**MOEA/D-AV**	MOEA/D	AWA	**MOEA/D-AV**
DTLZ1	0.97 (0.02)	0.98 (0.00)	0.95 (0.06)	0.21 (0.27)	0.25 (0.33)	0.51 (0.42)
DTLZ2	0.47 (0.00)	0.47 (0.00)	0.47 (0.00)	0.45 (0.00)	0.45 (0.00)	0.47 (0.00)
DTLZ3	0.45 (0.01)	0.45 (0.01)	0.44 (0.02)	0.01 (0.03)	0.02 (0.04)	0.15 (0.16)
DTLZ4	0.46 (0.00)	0.46 (0.00)	0.46 (0.00)	0.33 (0.12)	0.33 (0.15)	0.44 (0.05)
DTLZ5	0.46 (0.00)	0.22 (0.00)	0.22 (0.00)	0.22 (0.00)	0.22 (0.00)	0.22 (0.00)
DTLZ6	0.22 (0.00)	0.22 (0.00)	0.22 (0.00)	0.19 (0.08)	0.22 (0.00)	0.22 (0.00)
DTLZ7	0.24 (0.00)	0.24 (0.00)	0.24 (0.00)	0.13 (0.03)	0.21 (0.03)	0.22 (0.04)
ZDT1	0.66 (0.00)	0.66 (0.00)	0.67 (0.00)	0.03 (0.03)	0.03 (0.02)	0.12 (0.05)
ZDT2	0.33 (0.00)	0.33 (0.00)	0.33 (0.00)	0.00 (0.00)	0.00 (0.00)	0.00 (0.00)
ZDT3	1.04 (0.00)	1.04 (0.00)	1.04 (0.00)	0.20 (0.05)	0.23 (0.04)	0.33 (0.05)
ZDT4	0.66 (0.00)	0.66 (0.00)	0.66 (0.00)	0.00 (0.00)	0.00 (0.00)	0.00 (0.00)
ZDT6	0.33 (0.00)	0.33 (0.00)	0.33 (0.00)	0.00 (0.00)	0.00 (0.00)	0.00 (0.00)
Stats	=	=		+	+	
IGD	MOEA/D	AWA	**MOEA/D-AV**	MOEA/D	AWA	**MOEA/D-AV**
DTLZ1	0.58 (0.13)	0.58 (0.17)	0.57 (0.19)	1.33 (0.92)	4.38 (6.26)	0.96 (1.05)
DTLZ2	0.01 (0.00)	0.01 (0.00)	0.00 (0.00)	0.01 (0.00)	0.01 (0.00)	0.01 (0.00)
DTLZ3	0.01 (0.01)	0.01 (0.00)	0.02 (0.02)	4.98 (5.45)	6.96 (7.86)	2.14 (2.85)
DTLZ4	0.01 (0.00)	0.01 (0.00)	0.01 (0.00)	0.33 (0.31)	0.29 (0.34)	0.06 (0.16)
DTLZ5	0.00 (0.00)	0.00 (0.00)	0.00 (0.00)	0.00 (0.00)	0.00 (0.00)	0.00 (0.00)
DTLZ6	0.00 (0.00)	0.00 (0.00)	0.00 (0.00)	0.14 (0.34)	0.00 (0.00)	0.00 (0.00)
DTLZ7	0.01 (0.00)	0.01 (0.00)	0.01 (0.00)	0.39 (0.17)	0.13 (0.17)	0.15 (0.27)
ZDT1	0.00 (0.00)	0.00 (0.00)	0.00 (0.00)	0.73 (0.13)	0.69 (0.07)	0.50 (0.09)
ZDT2	0.00 (0.00)	0.00 (0.00)	0.00 (0.00)	1.29 (0.14)	1.18 (0.19)	0.91 (0.18)
ZDT3	0.00 (0.00)	0.00 (0.00)	0.00 (0.00)	0.61 (0.08)	0.57 (0.06)	0.46 (0.05)
ZDT4	0.00 (0.00)	0.00 (0.00)	0.00 (0.00)	21.1 (4.34)	20.5 (4.33)	17.5 (3.16)
ZDT6	0.00 (0.00)	0.00 (0.00)	0.00 (0.00)	5.64 (0.27)	5.55 (0.36)	5.02 (0.51)
Stats	=	=		+	+	
Entropy	MOEA/D	AWA	**MOEA/D-AV**	MOEA/D	AWA	**MOEA/D-AV**
DTLZ1	11.2 (0.88)	11.4 (0.73)	10.3 (1.15)	3.75 (0.45)	3.74 (0.59)	7.28 (2.04)
DTLZ2	12.2 (0.02)	12.3 (0.03)	12.2 (0.04)	9.19 (0.16)	8.62 (0.45)	11.1 (0.31)
DTLZ3	11.0 (0.72)	11.2 (0.58)	10.1 (1.28)	2.67 (0.48)	1.83 (0.88)	5.08 (2.58)
DTLZ4	11.1 (0.16)	11.1 (0.17)	10.9 (0.11)	6.00 (1.49)	5.47 (2.35)	8.32 (0.84)
DTLZ5	6.98 (0.03)	6.98 (0.02)	6.92 (0.02)	6.45 (0.05)	5.71 (0.24)	6.66 (0.03)
DTLZ6	6.58 (0.03)	6.59 (0.00)	6.60 (0.02)	6.44 (0.31)	5.29 (0.24)	6.17 (0.08)
DTLZ7	10.1 (0.08)	10.1 (0.14)	10.2 (0.25)	8.79 (0.91)	8.04 (0.70)	9.22 (0.44)
ZDT1	7.26 (0.03)	7.25 (0.03)	7.12 (0.22)	6.02 (0.27)	5.91 (0.28)	6.19 (0.20)
ZDT2	7.24 (0.04)	7.25 (0.09)	7.18 (0.17)	2.04 (1.50)	2.45 (1.47)	3.38 (1.41)
ZDT3	6.66 (0.06)	6.65 (0.05)	6.56 (0.50)	5.96 (1.60)	6.02 (0.18)	6.11 (0.11)
ZDT4	7.20 (0.06)	7.26 (0.02)	7.27 (0.08)	2.67 (0.54)	6.78 (0.44)	6.98 (0.25)
ZDT6	6.62 (0.24)	6.71 (0.29)	4.30 (0.37)	3.93 (0.47)	4.09 (0.61)	2.64 (1.07)
Stats	−	−	+	+		

lower number of weight vectors, the reason for its bad performance may be that such configuration provides little information about the search progress.

Figure 1 shows the change in the HV of the solutions of the final UEA for each method, on the DTLZ3 with continuous PF (a - left side) and DTLZ7 with discontinuous PF, (b - right side). On DTLZ3, MOEA/D-AV achieves higher or

Table 2. Paring of initial number of weight vectors and MOEA/D variant used for comparison of the "Best Scenario" and "Worst Scenario" (Table 1). These were selected using the best and worst mean HV values, respectively.

Best Scenario—Worst scenario						
	MOEA/D	AWA	MOEA/D-AV	MOEA/D	AWA	MOEA/D-AV
DTLZ1	190	105	190	990	990	990
DTLZ2	190	190	10	10	10	990
DTLZ3	190	105	190	990	990	990
DTLZ4	496	496	990	10	10	10
DTLZ5	190	105	21	21	990	990
DTLZ6	496	496	190	10	10	990
DTLZ7	496	190	10	10	10	21
ZDT1	50	10	100	1000	1000	1000
ZDT2	10	10	20	1000	1000	1000
ZDT3	50	50	10	1000	1000	1000
ZDT4	100	100	20	500	500	1000
ZDT6	50	50	100	500	1000	1000

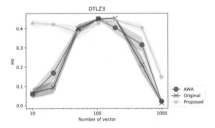
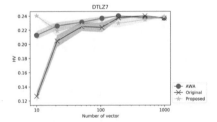

(a) MOEA/D-AV method finds better HV values in most initial settings. We highlight the results of MOEA/D-AV with only 10 initial vectors.

(b) MOEA/D-AV has the best performance at lower number of initial weight vectors and has competitive results for all settings.

Fig. 1. Mean HV value against initial number of weight vectors, for DTLZ3 on the left and DTLZ7 on the right. Shaded areas indicate standard deviations.

competitive results independently of the initial number of vectors and the performance of both MOEA/D-DE and MOEA/D-AWA deteriorates significantly when the number of weight vectors is not set correctly.

Figure 2 depicts the PF approximated for each method on DTLZ3. When the number of weight vectors is small, only MOEA/D-AV can provide a suitable approximation to the PF. We believe that reason why the other algorithms behave badly in this case is because the distance of the weight vectors provides little useful information about the search progress. This result supports the need

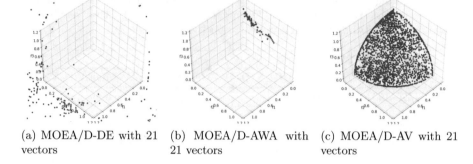

(a) MOEA/D-DE with 21 vectors

(b) MOEA/D-AWA with 21 vectors

(c) MOEA/D-AV with 21 vectors

Fig. 2. UEA of 3 methods starting from 21 vectors in DTLZ3. MOEA/D-DE and MOEA/D-AWA have low coverage of the PF, while MOEA/D-AV is able to cover well most regions of the objective space.

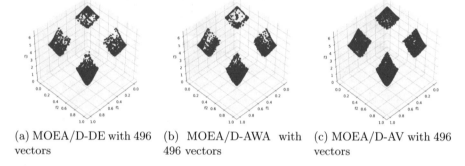

(a) MOEA/D-DE with 496 vectors

(b) MOEA/D-AWA with 496 vectors

(c) MOEA/D-AV with 496 vectors

Fig. 3. UEA of the 3 methods starting from 105 vectors in DTLZ7. Although there is little difference in the HV values for each of the methods, we can clearly see that MOEA/D-AV can fill empty regions of the objective space evenly.

to add vectors randomly to avoid early stagnation of the search, a feature only present in MOEA/D-AV.

Coming back to Fig. 1, we discuss the results of all algorithms in DTLZ7. MOEA/-AV is the only algorithm to achieve good results independently of the number of vectors. Although the HV performance of the methods is similar, Fig. 3 shows that their ability to cover the PF differs. MOEA/D-AV method provides a wider coverage of the optimal PF and this is related to the ability of this algorithm to add vectors to empty areas of the objective space and to remove weight vectors in areas of the PF already covered.

Figure 4 shows the change in the number of weight vectors for the best and worst setting scenario in DTLZ1 and ZDT2, respectively. At Fig. 4 (a - left side) the blue, continuous line illustrates that MOEA/D-AV to keep reducing and increasing the number of weight vectors until around 50000 evaluations. Then, the algorithm seems to keep increasing the weight vectors. On the other hand, Fig. 4 (b - right side) shows that MOEA/D-AV improves its coverage of the PF

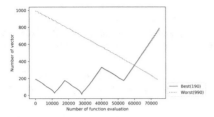
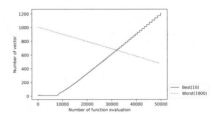

(a) Change in the number of vectors of MOEA/D-AV in DTLZ1. Best setting (blue, continuous line) starts with 190 vectors and the worst setting (brown, dashed line) starts with 990 vectors.

(b) Change in the number of vectors of MOEA/D-AV in ZDT2. Best setting (blue, continuous line) starts with 10 vectors and the worst setting (brown, dashed line) starts with 1000 vectors.

Fig. 4. MOEA/D-AV works better with low number of weight vectors, while higher values cause MOEA/D-AV to reduce the number of vectors.

a little earlier than before, at 10000 evaluations. In both cases the worst case keeps reducing the number of weight vectors, confirming that starting with high number of vectors deteriorates the performance of any MOEA/D.

6   Conclusion

Here we study the effect of adapting the number of weight vectors by removing unnecessary vectors and adding new vectors in empty areas of the PF. We proposed an the MOEA/D-AV that adaptively changes the number of weight vectors. MOEA/D-AV detects when the number of vectors must be changed and generates new vectors depending on the optimisation stage.

MOEA/D-AV has competitive performance, independently of the number of initial vectors. This result confirms that MOEA/D-AV finds suitable PF even when the number of initial vectors is not appropriate. One of the more significant findings is that using this method allows the use of MOEA/Ds without any fine-tuning process to choose the best set of the initial number of weight vectors. We understand that the dynamic adaptation of the number of weight vectors is a finding of interest for the whole Multi-Objective community. Future works include applying MOEA/D-AV in real-world MOPs, especially with non-regular and inverted PFs and comparing MOEA/D-AV and MOEA/D with Resource Allocation methods that activate and deactivate vectors during the search [8].

References

1. Črepinšek, M., Liu, S.H., Mernik, M.: Exploration and exploitation in evolutionary algorithms: a survey. ACM Comput. Surv. (CSUR) **45**(3), 1–33 (2013)
2. Deb, K., Thiele, L., Laumanns, M., Zitzler, E.: Scalable multi-objective optimization test problems. In: Proceedings of the 2002 Congress on Evolutionary Computation. CEC 2002 (Cat. No.02TH8600), vol. 1, pp. 825–830, (2002). https://doi.org/10.1109/CEC.2002.1007032

3. Farhang-Mehr, A., Azarm, S.: Diversity assessment of Pareto optimal solution sets: an entropy approach. In: Proceedings of the 2002 Congress on Evolutionary Computation. CEC 2002 (Cat. No.02TH8600), vol. 1, pp. 723–728. (2002). https://doi.org/10.1109/CEC.2002.1007015

4. de Farias, L.R.C., Braga, P.H.M., Bassani, H.F., Araújo, A.F.R.: MOEA/D with uniformly randomly adaptive weights. In: Proceedings of the Genetic and Evolutionary Computation Conference, pp. 641–648. GECCO 2018, Association for Computing Machinery, New York, NY, USA (2018). https://doi.org/10.1145/3205455.3205648

5. Glasmachers, T., Naujoks, B., Rudolph, G.: Start small, grow big? Saving multi-objective function evaluations. In: Bartz-Beielstein, T., Branke, J., Filipič, B., Smith, J. (eds.) PPSN 2014. LNCS, vol. 8672, pp. 579–588. Springer, Cham (2014). https://doi.org/10.1007/978-3-319-10762-2_57

6. Goel, T., Stander, N.: Non-dominance-based online stopping criterion for multi-objective evolutionary algorithms. Int. J. Numer. Meth. Eng. **88**, 661–684 (2010)

7. Jiang, S., et al.: Towards adaptive weight vectors for multiobjective evolutionary algorithm based on decomposition. In: 2016 IEEE Congress on Evolutionary Computation (CEC), pp. 500–507 (2016). https://doi.org/10.1109/CEC.2016.7743835

8. Lavinas, Y., Aranha, C., Ladeira, M., Campelo, F.: MOEA/D with random partial update strategy. In: 2020 IEEE Congress on Evolutionary Computation (CEC), pp. 1–8 (2020)

9. Li, H., Zhang, Q.: Multiobjective optimization problems with complicated pareto sets, MOEA/D and NSGA-II. IEEE Trans. Evol. Comput. **13**(2), 284–302 (2009). https://doi.org/10.1109/TEVC.2008.925798

10. Li, M., Yao, X.: What weights work for you? Adapting weights for any pareto front shape in decomposition-based evolutionary multiobjective optimisation. Evol. Comput. **28**(2), 227–253 (2020)

11. Ma, X., Yu, Y., Li, X., Qi, Y., Zhu, Z.: A survey of weight vector adjustment methods for decomposition based multi-objective evolutionary algorithms. Evol. Comput. **24**(4), 634–649 (2020)

12. Qi, Y., Ma, X., Liu, F., Jiao, L., Sun, J., We, J.: MOEA/D with adaptive weight adjustment. Evol. Comput. **22**(2), 231–264 (2014)

13. Tanabe, R., Ishibuchi, H., Oyama, A.: Benchmarking multi- and many-objective evolutionary algorithms under two optimization scenarios. IEEE Access **5**, 19597–19619 (2017)

14. Wagner, T., Trautmann, H., Naujoks, B.: OCD: online convergence detection for evolutionary multi-objective algorithms based on statistical testing. In: Ehrgott, M., Fonseca, C.M., Gandibleux, X., Hao, J.-K., Sevaux, M. (eds.) EMO 2009. LNCS, vol. 5467, pp. 198–215. Springer, Heidelberg (2009). https://doi.org/10.1007/978-3-642-01020-0_19

15. Zhang, Q., Li, H.: MOEA/D: a multiobjective evolutionary algorithm based on decomposition. IEEE Trans. Evol. Comput. **11**(6), 712–731 (2007)

16. Zitzler, E., Deb, K., Thiele, L.: Comparison of multiobjective evolutionary algorithms: empirical results. Evol. Comput. **8**(2), 173–195 (2000)

Parallel Asynchronous Memetic Optimization for Freeform Optical Design

Maxim Sakharov[1]([✉]) [ID], Thomas Houllier[2,3] [ID], and Thierry Lépine[2,4] [ID]

[1] Bauman Moscow State Technical University, 5/1 2-ya Baumanskaya, 105005 Moscow, Russia
[2] Univ-Lyon, Laboratoire Hubert Curien, UMR CNRS 5516, 18 rue Benoît Lauras, 42000 Saint-Etienne, France
[3] Sophia Engineering, 5 Rue Soutrane, 06560 Sophia Antipolis, France
[4] Institut d'Optique Graduate School, 18 rue Benoît Lauras, 42000 Saint-Etienne, France

Abstract. Freeform optical systems include lens or mirror surfaces with shapes that have no axis of revolution in order to reach better performance in off-axis designs such as Head-Mounted Displays (*HMD*). They have an increased number of design parameters compared to rotationally symmetrical optical systems which makes their optimization more difficult. In this paper we presented results of several studies where various optimization techniques were used to optimize a freeform *HMD* prism with two mirrors. It was demonstrated that algorithms which performed well on traditional systems were not capable of obtaining high-quality solutions for the system under investigation. This paper presents a new parallel asynchronous memetic algorithm designed for optimizing freeform optical systems. The algorithm incorporates local search techniques into an asynchronous parallel computing procedure thus helping to speed-up the convergence to a high-quality solution. The results of a comparative study, presented in this paper along with the algorithm description, demonstrate that the proposed method is not only capable of obtaining a good solution but is also comparable with high-end commercial optical software packages.

Keywords: Global optimization · Freeform optics · Memetic algorithms · Parallel computing

1 Introduction

The design of optical systems often involves optimization procedures to determine the geometrical parameters of lens or mirror surfaces in order to obtain the best possible optical properties of a system. This is why several optimization algorithms are typically included in optical design software. As a rule, optical designers define a scalar objective that comprises all the optical quality and geometrical constraints specifications and optimize the system parameters. Optical system parameters usually are radii of curvature of lenses/mirrors, relative geometrical position of optical elements in the system, choice of glass for the lenses, lists of surface parameters for more complex optical surfaces, etc.

In the field of optical design many researchers put significant effort into studying different optimization methods and their performance [1–3]. What is more, the emerging

© Springer Nature Switzerland AG 2021
C. C. Aranha et al. (Eds.): TPNC 2021, LNCS 13082, pp. 97–108, 2021.
https://doi.org/10.1007/978-3-030-90425-8_8

field of freeform optics, which commonly designates the study of optical systems with optical surfaces that do not have an axis of rotational symmetry, allows the design of very compact, very performant systems [4, 5]. The use of freeform surfaces creates optimization problems with more degrees of freedom than in conventional optics. This makes the optimization process more difficult. The problem has prompted an interest in investigating various mathematical representations for freeform surfaces in their interplay with the search process [6, 7]. The objective function and constraints in such optimization problems often show non-linear behavior and presence of many local minima and the number of input parameters can range from dozens to several hundreds. This eventually creates challenges for optimization methods.

As it was demonstrated in [8–10], any optimization method in order to be successful not for a specific problem but at least for a certain class of problems, must possess some adaptation properties. Literature shows various approaches to adaptation which are used in practice. This includes initial data analysis, dimensionality reduction of a search domain, landscape analysis of the objective function, hybridization with local or global search methods, co-modification of algorithms, etc. [11, 12]. The methodological background for these techniques is the *No Free Lunch* theorem [13] that implies the following. If algorithm A_1 is more efficient for solving a particular problem than algorithm A_2, then another problem exists where algorithm A_2 will outperform algorithm A_1.

In this work we continue our research in the field of optimization of optical systems. We present results of several studies where various optimization techniques were used to optimize a freeform *HMD* prism with two mirrors [14]. It was demonstrated that algorithms which performed well on conventional systems [15] were not capable of obtaining high-quality solutions for the freeform system under investigation. This paper also presents a new parallel asynchronous memetic algorithm designed for optimizing freeform optical systems. The algorithm incorporates local search techniques into an asynchronous parallel computing procedure thus helping to speed-up the convergence to a high-quality solution.

The rest of the paper is organized as follows. Section 2 is devoted to a problem formulation. Section 3 contains a description of the freeform optical system under investigation as well as the results of previous studies which involved both commercial and freely available optimization methods. In Sect. 4 we present the asynchronous memetic algorithm designed to cope with the specified optimization challenges. Sections 5 and 6 contain the description conducted numerical experiments and the analysis of obtained both from optimization and optical points of view. Section Conclusion summarizes the study and points directions for further work.

2   Problem Formulation

We consider a deterministic global constrained minimization problem

$$\min_{X \in D} \Phi(X) = \Phi(X^*) = \Phi^*. \tag{1}$$

Here $\Phi(X)$ is the scalar objective function, $\Phi(X^*) = \Phi^*$ is its required minimal value, $X = (x_1, x_2, \ldots, x_{|X|}) - |X|$-dimensional vector of variables.

Feasible domain D is determined with inequality constraints

$$D = \left\{ X \mid x_i^{min} \leq x_i \leq x_i^{max}, i \in [1 : |X|] \right\} \subset R^{|X|}. \tag{2}$$

In optical system design, the quality of a system can be measured using a so-called merit function (*MF*). The goal is to minimize the value of *MF*: a lesser value of *MF* indicates a better optical design, but at the same time *MF* cannot be less than zero. In this work we use the same merit function that was defined by the authors in [14]. This function takes two criteria into account.

- *Spot size*. Each bundle of rays coming from a given field at the entrance of the optical system must form the smallest possible spot on the image plane. This is an image quality criterion, if it is low then the image will be sharp, otherwise it will be blurry
- *Image position*. Each bundle of rays coming from a given field at the entrance of the optical system must focus on a given position on the image plane. This is a control over the focal length as well as the distortion in the optical system

Our definition of *MF* is used as the objective function $\Phi(X)$ and can be described as follows

$$\Phi(X) = \frac{\sum_{f=1}^{N_{fields}} Spot_f + \alpha Pos_f}{N_{fields}}. \tag{3}$$

Here N_{fields} is the number field points used for computation (typically up to a dozen), $Spot_f$ is the spot size criterion for field f, Pos_f is the image position criterion for field f, α is the weighting coefficient.

3 Freeform Optical Design

An application of freeform optics is reaching better imaging performance, and/or smaller system size and mass for systems that are off-axis. Examples of such off-axis systems are unobscured space telescopes and instruments or prisms for *HMD* [16].

The system we investigate in this paper, presented in Fig. 1, is a freeform prism with two optical surfaces described by the fifth order XY polynomials. The whole prism is plano-symmetric. This system has 22 degrees of freedom. The system is inspired from the general shape of the system in [4]. The aperture stop diameter is the diameter of the eye pupil. The focal length is constrained by the field of view in the eye space and size of the object. The raytracing for this system is done at a single wavelength.

The vector of input variables X for the specified system is made of 22 components: x_1 and x_2 represent the curvatures for S2 and S3 correspondingly; x_3 through x_{12} represent polynomial coefficients for S2; x_{13} through x_{22} do for S3. Bounds for each variable are listed in Table 1. These values identify geometric properties of an optical system and allows evaluating its integral optical properties based on the value of the merit function.

In [14] the authors have conducted an extensive empirical investigation of different derivative-free optimization methods. The study involved the development of the following methods: the Particle Swarm Optimization algorithm (*PSO*) [17], The Gravity

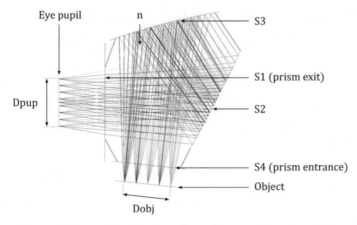

Fig. 1. A layout of the freeform prism with two mirrors (S2 and S3)

Table 1. Constraints for input variables of the freeform optical system.

Parameter	Constraints
Curvature	$[-0.1; 0.1]\,\text{mm}^{-1}$
XY coefficients	$[-0.5; 0.5]\,\text{mm}$

Search Algorithm (*GSA*) [18], the Cuckoo Search algorithm (*CS*) [19], the Covariance Matrix Adaptation Evolution Strategy (*CMAES*) [20], the Nelder-Mead algorithm (*SPX*) [21], and the simple Monte-Carlo algorithm [17]. In addition, the system was recreated and optimized in *Zemax OpticStudio* which is one of the most powerful software packages for optics nowadays. *OpticStudio* provides several built-in optimization methods; two of them were used in our study, namely, Local optimization Orthogonal Descent (*ZLOD*) and Hammer Orthogonal Descent (*ZHOD*). All values of the algorithms' free parameters are described in detail in [14].

As a part of our current work, we extended that empirical study with *CoMEC* algorithm introduced by the authors in [15]. This algorithm is a co-modification of the traditional Mind Evolutionary Computation algorithm [22]. It demonstrated very good results when optimizing conventional optical systems [15].

Each optimization method was launched 100 times with random initial conditions. The computational budget was limited to 5000 evaluations of the objective function. Figure 2 presents the obtained results in logarithmic scale: x-axis represents the values of the objective function while y-axis does the number of launches. Every curve demonstrates the increasing (from left to right) number of launches (N out of 100) that produced the *MF* values equal or less than the specific value of x. For instance, nearly 90 runs of the *PSO* algorithm produced the objective function's value that is less than 0.01 which we denote as threshold Δf (thick vertical line in Fig. 2). Since the exact global minimum for this system is unknown, we use the threshold to distinguish between acceptable solutions and not acceptable ones.

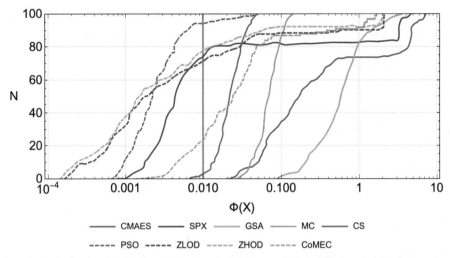

Fig. 2. Optimization results for different optimization methods applied to the freeform prism system

Obtained results showed that commercial optimization methods appeared to be most efficient for the system under consideration in terms of least *MF* values. Regarding the number of successful runs, their performance was similar to the Nelder-Mead algorithm, while all of them were inferior to the *PSO* method. Results of the *CoMEC* method for this system turned out to be quite moderate providing only 20 successful runs despite being very efficient for conventional optical systems. This result agrees with the No Free Lunch theorem [13].

4   Memetic Asynchronous MEC Algorithm

The *CoMEC* algorithm was introduced and described in detail in [15]. The results presented in the previous section demonstrated that this algorithm lacks the convergence speed to provide sufficient results under the computational budget of 5000 evaluations of the objective function.

In order to overcome this disadvantage, we propose an asynchronous memetic modification of the canonical *MEC* algorithm [22] that was used as a starting point for the *CoMEC* algorithm. Memetic algorithms (*MA*) represent population meta-heuristic optimization algorithms based on the neo-Darwinian evolution and a concept of meme as a unit of cultural information proposed by R. Dawkins [23] to describe the learning process. In the context of *MA*, a meme can be considered as any optimization method applied to a current solution during the evolution process. This class of algorithms proved to be successful for different practical application [24, 25].

The idea of asynchronous learning was inspired by the work [26] where it was used for deep evolutionary reinforcement learning for large-scale problems. Typical evolutionary simulations employed generational evolution, where at each generation, the entire population is simultaneously replaced by applying various evolutionary operators to the fittest individuals. This paradigm can be slow due to the significant computational burden imposed by optimizing every member of a large population before any further evolution can occur.

The original *MEC* algorithm [22] is inspired by a human society and simulates some aspects of human behavior. An individual *s* is considered as an intelligent agent which operates in a group *S* made of analogous individuals. During the evolution process every individual is affected by other individuals within a group. In order to achieve a high position within a group, an individual has to learn from the most successful individuals in this group, while groups should follow the same logic in the intergroup competition.

Canonical *MEC* is composed of three main stages: initialization of groups, similar-taxis and dissimilation. Operations of similar-taxis and dissimilation are repeated iteratively while the best obtained value of an objective function $\Phi(X)$ is changing. When the best obtained value stops changing, the winner of the best group from a set of leading ones is selected as a solution to the optimization problem [17].

The proposed Memetic Asynchronous MEC algorithm (*MAMEC*) belongs to a class of parallel algorithms and utilizes the master-slave parallelization paradigm [24]. It allows moving from generational to asynchronous parallel evolution, where it is not required for learning to be finished across the entire population before any further evolution occurs. Instead, as soon as any agent finishes learning, the worker (CPU instance) can immediately perform another step of the *MEC* algorithm. This would help improving the convergence speed and obtaining a sufficient solution within the defined computational limit.

The detailed description of the *MEC* algorithm is presented in [25]. In this paper we provide a brief description of the *MAMEC* algorithm focusing only on those parts that are different from the canonical *MEC*.

1. Initialization of groups within the search domain *D*.

   (a) Generate a given number γ of groups S_i, $i \in [1 : \gamma]$; γ is the free parameter of the algorithm.
   (b) Generate a random vector $X_{i,1}$. Identify this vector with the individual $s_{i,1}$ of the group S_i.
   (c) Determine the initial coordinates of the rest of individuals in the group $S_{i,j}, j \in [2 : |S|]$ according to the formula

   $$X_{i,j} = X_{i,1} + N_{|X|}(0, \sigma). \tag{4}$$

   (d) Calculate the scores of all individuals in every population S_i and put them on the corresponding local blackboards.
   (e) Create leading S^b and lagging S^w groups based on the obtained scores.

2. The modified local search stage is launched independently on every parallel computing node and works with a random subset of *K* groups. We describe the process for one CPU instance.

   (a) Take information on the current best individual $s_{i,j}, j \in [1 : |S_i|]$ of the group S_i from the blackboard C_i.

(b) Apply a local search method to the best individual $s_{i,k}$. In this work the Hooke-Jeeves method [17] was used as a meme since this is a deterministic gradient-free method.

(c) Take new best individual $\tilde{s}_{i,k}$ and compose a new group using formula (4).

(d) Put information on the new winner on the corresponding local and global black-boards. Global blackboard is being updated by every CPU worker after the learning process is completed. The information from this blackboard is used at the dissimilation stage to eliminate unsuccessful groups and save computational recourses.

3. The dissimilation stage is performed by the master CPU after every global blackboard update by the workers.

(a) Read the scores of all groups $\Phi_i^b, \Phi_j^w, i \in \left[1 : \left|S^b\right|\right], j \in [1 : |S^w|]$ from the global blackboard C^g.

(b) Compare those scores. If score of any leading group S_i^b appeared to be less than score of any lagging group S_j^w, than the latter becomes a leading group, and the first group becomes a lagging one. If score of a lagging group S_k^w is lower than scores of all leading groups for ω consecutive updates, then it's removed from the population.

(c) Using the initialization operation each removed group is replaced with a new one.

(d) If any lagging group S_j^w was removed while being processed on another computing node, its results are kept to form a new group as soon as another group is removed. It provides the constant number of groups and individuals and prevents information loss.

4. As the termination criteria the maximum allowed number of the objective function's evaluation was used.

5 Numerical Experiments

The *MAMEC* algorithm was implemented by the authors in *Wolfram Mathematica*. Software implementation has a modular structure, which helps to use different local search techniques easily. *Mathematica* also includes tools to manage parallel computing nodes. The proposed algorithm and its software implementation were used to optimize the specified freeform system in the same manner as in previous studies [14]. The two-mirror system was built in *Zemax OpticStudio* which was utilized to calculate the objective function. The integration between *OpticStudio* and *Mathematica* was set up using *Zemax API* so that at every step of the *MAMEC* algorithm a new vector X is sent to *OpticStudio* for calculating the objective function's value and after that this value is sent back to *Mathematica* for subsequent operations.

The *MAMEC* algorithms was launched $N = 100$ times with random initial conditions with the same computational budget of 5000 evaluations of the objective function. The obtained results presented in Fig. 3 demonstrate that the *MAMEC* algorithm provides much better results than the *CoMEC* algorithm and several other optimization methods. Among all methods under investigation the *MAMEC* algorithms provides the highest fraction of feasible solutions (solutions that are below the specified threshold $\Delta f = 10^{-2}$). It is also worth mentioning that best 30% of obtained solutions are relatively close to the best ones obtained by commercial software. These results are promising for further improvements of the *MAMEC* algorithm.

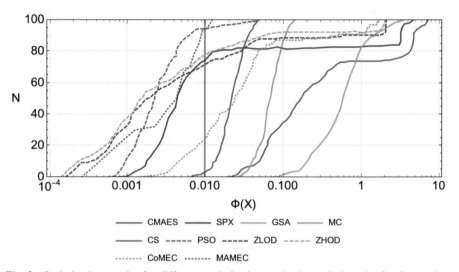

Fig. 3. Optimization results for different optimization methods applied to the freeform prism system including the proposed *MAMEC* algorithm

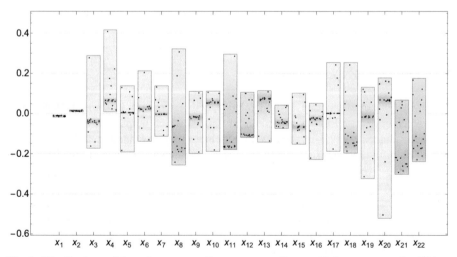

Fig. 4. Distributions of the values across all components of vector X that correspond to 20 best solutions obtained with a use of the *MAMEC* algorithm

To better understand the search space and possible challenges we investigated 20 best solutions found with a use of the *MAMEC* algorithm. Distributions across all components of vector X for 20 best solutions are presented in Fig. 4. This graph suggests a high number of local optima. While for all solutions values of curvature for two surfaces (x_1 and x_2) are almost identical we can see wide spreads for XY coefficients (x_3 through x_{22}). For many variables those spreads are formed with a few outliers and dense areas of solutions are clearly seen (for example, x_3 and x_4). This can be an indicator of premature convergence of the algorithm. However, for several variables (such as x_8, x_{11}, and x_{21}) we can see that there is no clear dense area, which can be an indicator of many different local minima in the cross-section of such a variable. What is more, variables x_{21} and x_{22} can be either positive or negative and still contribute to a high-quality overall solution.

6   Analysis from the Optical Perspective

In this section we analyze several best system configurations found by the *MAMEC* algorithm from the perspective of an optical designer. We report that although the solution vectors seem different, as shown above, the resulting optical performance of each solution is quite similar. As a preliminary observation, this seems to suggest the problem is degenerate. We present a typical result in Fig. 5 (spot diagrams) and Fig. 6 (grid distortion). The figures are obtained using *Zemax* optical software. The spot diagrams are the impact of rays on the system image plane at a few points across the field of view. Diagrams with points clustered closer together are characteristic of an optical system with good imaging performance (the image will be sharp). In the context of *HMD* this

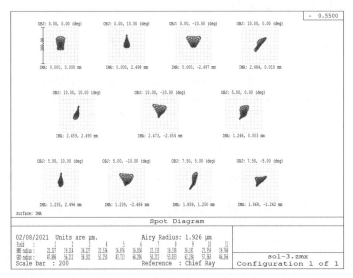

Fig. 5. Spot diagrams for a typical high-quality result given by the *MAMEC* algorithm. RMS radii of spots are approximately 20 microns

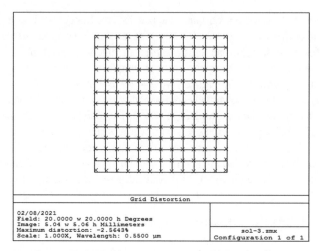

Fig. 6. Grid distortion in one of the obtained solutions

means we can use displays with smaller pixels (and thus higher resolution). The distortion grid shows the geometric distortion of the image as it passes through the optical system.

The optical performance of the obtained systems is sufficient to serve as good starting point for a final tuning by optical design experts. The fine-tuning process would typically involve, in the present case, relaxing the distortion constraint and strengthening the spot size constraint, as the system shown here has poor image sharpness but very good distortion performance for an *HMD*.

7 Conclusions

This paper presents the memetic asynchronous *MEC* algorithm designed for optimizing freeform optical systems with a relatively small computational budget. Instead of a traditional generational paradigm, the *MAMEC* algorithm operates according to an asynchronous parallel evolution model, where it is not required for learning to be finished across the entire population before any further operation. Once local optimization by a meme is completed for any individual, the master process can immediately perform another step of the basic algorithm. As a result, it provides a better solution for a smaller number of evaluations.

Results of the extensive comparative study of ten different optimization methods are presented in this work along with analysis of obtained systems from the perspective of an optical engineer. This analysis suggests that the optical performance of the obtained systems is good enough to be used as an initial system for fine tuning by the experts. The *MAMEC* algorithm is capable of finding high-quality solutions, that are comparable with the commercial optimizers included into the high-end optical design software.

Further research will be devoted to improving the *MAMEC* algorithm by incorporating a landscape analysis procedure for a preliminary analysis of optimization problems and studying its performance for other freeform optical systems. It is also worth investigating the algorithms' efficiency when using other local search methods or multi-memes which combine several optimization algorithms.

References

1. Shafer, D.: Global optimization in optical design. Comput. Phys. **8**(12), 188–195 (1994)
2. Höschel, K., Lakshminarayanan, V.: Genetic algorithms for lens design: a review. J. Opt. **48**(1), 134–144 (2018). https://doi.org/10.1007/s12596-018-0497-3
3. Cheng, X., Yongtian, W., Qun, H., Masaki, I.: Global and local optimization for optical systems. Optik **117**(13), 111–117 (2006)
4. Shen, Z., et al.: Customized design and efficient fabrication of two freeform aluminum mirrors by single point diamond turning technique. Appl. Opt. **58**(19), 2269–2276 (2019)
5. Meng, Q., Wang, H., Liang, W., Yan, Z., Wang, B.: Design of off-axis three-mirror systems with ultrawide field of view based on an expansion process of surface freeform and field of view. Appl. Opt. **58**(13), 609–615 (2019)
6. Yabe, A.: Representation of freeform surfaces suitable for optimization. Appl. Opt. **51**(115), 3054–3058 (2012)
7. Broemel, A., Liu, C., Zhong, Y., Zhang, Y., Gross, H.: Freeform surface descriptions. Part II: application benchmark. Adv. Opt. Technol. **6**(15), 337–347 (2017)
8. Neri, F., Cotta, C., Moscato, P.: Handbook of Memetic Algorithms, 368 p. Springer, Heidelberg (2011). https://doi.org/10.1007/978-3-642-23247-3
9. Sakharov, M., Karpenko, A.: Parallel multi-memetic global optimization algorithm for optimal control of polyarylenephthalide's thermally-stimulated luminescence. In: Le Thi, H.A., Le, H.M., Pham Dinh, T. (eds.) WCGO 2019. AISC, vol. 991, pp. 191–201. Springer, Cham (2020). https://doi.org/10.1007/978-3-030-21803-4_20
10. Sakharov, M., Karpenko, A.: Multi-memetic mind evolutionary computation algorithm based on the landscape analysis. In: Fagan, D., Martín-Vide, C., O'Neill, M., Vega-Rodríguez, M.A. (eds.) TPNC 2018. LNCS, vol. 11324, pp. 238–249. Springer, Cham (2018). https://doi.org/10.1007/978-3-030-04070-3_19
11. Weise, T.: Global optimization algorithms - theory and application, 758 p. University of Kassel (2008)
12. Mersmann, O., et al.: Exploratory landscape analysis. In: Proceedings of the 13th Annual Conference on Genetic and Evolutionary Computation, pp.829–836. ACM (2011). https://doi.org/10.1145/2001576.2001690
13. Wolpert, D.H., Macready, W.G.: No free lunch theorems for optimization. IEEE Trans. Evol. Comput. **1**(1), 67–82 (1997)
14. Houllier, T., Lépine, T.: Comparing optimization algorithms for conventional and freeform optical design. Opt. Express **27**, 18940–18957 (2019)
15. Sakharov, M., Houllier, T., Lépine, T.: Mind Evolutionary computation co-algorithm for optimizing optical systems. In: Kovalev, S., Tarassov, V., Snasel, V., Sukhanov, A. (eds.) IITI 2019. AISC, vol. 1156, pp. 476–486. Springer, Cham (2020). https://doi.org/10.1007/978-3-030-50097-9_48
16. Borguet, B., Moreau, V., Santandrea, S., Versluys, J., Bourdoux, A.: The challenges of broadband performances within a compact imaging spectrometer: the ELOIS solution. In: Proceedings of SPIE, International Conference on Space Optics—ICSO 2020, vol. 11852, p. 118521E (2021). https://doi.org/10.1117/12.2599239
17. Karpenko, A.P.: Modern algorithms of search engine optimization. Nature-inspired optimization algorithms, 446 p. Bauman MSTU Publ., Moscow (2014). (in Russian)

18. Rashedi, E., Hossein, N.-P., Saeid, S.: GSA: a gravitational search algorithm. Inf. Sci. **179**(113), 2232–2248 (2009)
19. Yang, X.-S., Suash, D.: Engineering optimisation by cuckoo search. Math. Modell. Numer. Optim. **1**(14), 330–343 (2010)
20. Hansen, N., Ostermeier, A.: Completely derandomized self-adaptation in evolution strategies. Evol. Comput. **9**(12), 159–195 (2001)
21. Nelder, J.A., Mead, R.: A simplex method for function minimization. Comput. J. **7**(14), 308–313 (1965)
22. Chengyi, S., Yan, S., Wanzhen, W.: A survey of MEC: 1998–2001. In: 2002 IEEE International Conference on Systems, Man and Cybernetics IEEE SMC2002, Hammamet, Tunisia, 6–9 October 2002, vol. 6, pp.445–453. Institute of Electrical and Electronics Engineers Inc. (2002). https://doi.org/10.1109/ICSMC.2002.1175629
23. Dawkins, R.: The Selfish Gene, 384 p. Oxford University Press (1976)
24. Sakharov, M.K.: New adaptive multi-memetic global optimization algorithm for loosely coupled systems. Herald of the Bauman Moscow State Technical University, Series Instrument Engineering, no. 5, pp. 95–114 (2019). (in Russian). https://doi.org/10.18698/0236-3933-2019-5-95-114
25. Sakharov, M.K., Karpenko, A.P.: Adaptive load balancing in the modified mind evolutionary computation algorithm. Supercomput. Front. Innov. **5**(4), 5–14 (2018). https://doi.org/10.14529/jsfi180401
26. Gupta, A., Savarese, S., Ganguli, S., Fei-Fei, L.: Embodied intelligence via learning and evolution. https://arxiv.org/abs/2102.02202. Accessed 03 Aug 2021

Ant-Based Generation Constructive Hyper-heuristics for the Movie Scene Scheduling Problem

Emilio Singh[ID] and Nelishia Pillay[(✉)][ID]

Department of Computer Science, University of Pretoria, Pretoria, South Africa
u14006512@tuks.co.za, npillay@cs.up.ac.za

Abstract. The task of generation constructive hyper-heuristics concerns itself with generating new heuristics for problem domains via some kind of mechanism that combines low-level heuristic components into new heuristics. The movie scene scheduling problem is a recently developed combinatorial problem for which there are relatively few low-level heuristics. This paper focused on the application of a novel ant-based generation constructive hyper-heuristic to develop new constructive heuristics for the problem. The ant-based generation constructive hyper-heuristic was applied to create components that were themselves produced from existing heuristics and domain knowledge regarding the movie scene scheduling problem. The results of the research demonstrated that the ant-based hyper-heuristic was successful in the domain. It outperformed the existing set of human-derived constructive heuristics across a wide variety of problem classes and over several instances within the movie scene scheduling problem. The success of this research suggests that other hyper-heuristic methods, such as a generation perturbative one, could be applied to the movie scene scheduling problem in the future.

Keywords: Generation constructive hyper-heuristics · Ant algorithms · Discrete combinatorial optimization

1 Introduction

The costs associated with movie production have risen as the complexities of movie production have increased [1]. With rising costs and growing production areas, it has become more important for studios to incorporate improvements to their production schedules to reduce the overall costs of movie production. From this desire, the movie scene scheduling problem (MSSP) has emerged as an NP-hard scheduling problem. It primarily considers the task of producing a scene order schedule that minimises the production costs of those scenes [3].

As a scheduling problem, the task considers how to order n scenes from a set S such that the production costs of these scenes is minimised. The production

© Springer Nature Switzerland AG 2021
C. C. Aranha et al. (Eds.): TPNC 2021, LNCS 13082, pp. 109–120, 2021.
https://doi.org/10.1007/978-3-030-90425-8_9

costs primarily entail the costs of transferring actors between scenes as well as wages paid to the actors for their time and work.

More recent research has considered the effect of applying hyper-heuristics to the problem [12]. Specifically, the authors applied selective constructive and perturbative hyper-heuristics. This left open the potential to more deeply study the problem by applying generation constructive hyper-heuristic techniques to it as a way of expanding research in a field where human-derived heuristics are generally unavailable owing to the problem's recency of development. A hyper-heuristic ant colony optimisation algorithm (HACO) method has already been successfully applied as a generation constructive hyper-heuristic to the 1D bin packing problem and travelling salesperson domains [10]. The MSSP domain remains open for a successful application.

Thus the main contribution of this paper is the novel application of the HACO algorithm to the MSSP domain where no generation constructive hyper-heuristics have been applied before. In particular, previous work has applied the HACO algorithm to existing and well-known benchmark problems. This application here is on a novel domain. The rest of the paper is structured as follows. In Sect. 2, a background to the problem is presented. Section 3 presents the definition of the MSSP. The HACO algorithm is presented in Sect. 4. The experimental procedure is explained in detail in Sect. 5. The results of the subsequent experiments are presented in Sect. 6. Finally, Sect. 7 provides a conclusion to the paper with an overview of the findings and avenues for future research.

2   Background

In this section, an overview of relevant topics is presented. This will provide a summary of the MSSP as well as relevant hyper-heuristic techniques as they apply.

The MSSP as a problem is related to a class of entertainment scheduling problems. One example of this would be the scheduling of musicians that rehearse music [4]. In that case, the musicians are similar to actors with their pieces of music being similar to scenes. Later researchers extended this to consider unequal length musical pieces with a two-stage model checking procedure [9].

Some of the earliest work that more directly involved scheduling movies made use of scenes of a fixed length [3]. The variable that influenced production costs in this was the variability of actor wages. Later authors applied a genetic algorithm to this problem with improved results [8].

Other authors attempted to increase the complexity of scene scheduling by considering scenes of varying durations in conjunction with varying actor wages [2]. For that case, dynamic programming was applied with some success.

The next step forward for the MSSP was the inclusion of different scene shooting locations which more closely mirrors the actual reality of movie production [5]. With this new consideration came the inclusion of additional costs such as the costs of transferring actors between locations and the time required to do it. Initially, tabu-search methods were used but this later included both particle swarm optimisation (PSO) and ant colony optimisation (ACO) methods [6].

More recently, the use of hyper-heuristics was considered for the MSSP [12]. The authors made use of an ant-based hyper-heuristic for both selection constructive and selection perturbative hyper-heuristics to some success. One area in particular that proved potentially problematic was the degree to which the human-derived heuristics were able to affect solutions. As the MSSP is a new problem, the development of a body of low-level heuristics remains largely incomplete.

Thus, it is apparent from the survey of existing research that there is clear research potential for the use of an ant-based hyper-heuristic to generate new heuristics for the MSSP. In particular, the focus is applied to the task of generation constructive heuristics because good solution construction methods are the foundation of future studies into the problem domain.

3 Movie Scene Scheduling Problem

In this section, the MSSP is formally defined with its mathematical model, parameters, and inputs. The definition presented here is based on the formulation that extended the model in prior work [12].

3.1 Problem Definition

The MSSP contains the following elements: the set of n scenes S, the set of m locations L, and the set of o actors A. All scenes must be scheduled only once with no scenes overlapping each other and every scene assigned to only one location. Scenes may be assigned to the same location but this must be at different times. At least one actor must be assigned to a scene although every actor can be in many non-current scenes.

There are several secondary variables required for defining the MSSP:

- W: This defines a set of daily actor wages. They are paid for each day of production they are involved in including downtime between two scenes. Each actor's daily wage is in the range of [50,100].
- D: This defines the duration of each scene in days. Every scene once scheduled will be fully completed. Each scene's duration is in the range of [1,10] days.
- O: This refers to the location assigned to scene i. Each scene is assigned a randomly determined location out of the list of locations L. Every location has an equal probability of selection and every scene must be assigned a location.
- T_{xy}: This variable is a matrix of transfer times (in days) between different locations. The transfer time to move from location x to x is 0. Each value is in the range [1,10] days.
- C_{xy}: This variable is a matrix of transfer costs between different scenes. The transfer cost to move from scene x to x is 0. Each value is in the range [100,999].
- AS: This quantifies the assignment of o actors to scene i. For each scene, a randomly shuffled list of

all of the actors is generated. A random number of actors, in the range of $[1,n_s]$, are removed from this list. The remaining actors are then assigned to scene i.

The task of MSSP, therefore, is to order the set S into a schedule R such that the costs of scheduling the scenes are minimised.

4   Hyper-heuristic Ant Colony Optimisation

The high-level algorithm of the HACO method is presented in Algorithm 1. This algorithm details a broad overview of how the algorithm functions. As this research focuses on the application of the HACO algorithm to the MSSP, a full explanation of the HACO algorithm can be found in prior research [10,11].

Algorithm 1: High Level Algorithm

Input: n_k ant colony, it the max number of iterations, ph a pheromone map, p the evaporation rate, α the pheromone desirability, $size_d$ the path limit

Result: S_B the best solution, P_B the best path

1 initialise ph with small random values;
2 **foreach** $a \in n_k$ **do**
3   initialise a

4 $i=0$;
5 **for** $i < it$ **do**
6      $fitness[] = [n_k]$;
7      $best = \inf$;
8   **foreach** $a \in n_k$ **do**
9   construct a path
10   **foreach** $ant\ a\ in\ n_k$ **do**
11          $fitness[a]$=evaluate(a);
12   **if** $fitness[a] < best$ **then**
13              $best = fitness[a]$;
14   S_B=a.getSolution();
15   P_B=a.getPath();

16   evaporate ph update ph adjust p adjust α
17   **if** $fitness\ stagnation\ detected$ **then**
18   break;
19      $i = i + 1$;

The algorithm starts with a population of ants initialised with empty paths (lines 2 and 3). These ants will construct paths, line 9, over some iterations with each iteration producing a complete path through the component space. A fitness function is used to evaluate the path (line 11) and the pheromone map (lines 16 and 17) will be updated. The output of the algorithm after the iterations is the best solution found as well as the best path.

The algorithm uses the number of iterations as the main stopping condition. If, however, a fitness plateau is reached (that is fitness stagnation) during the run then the algorithm will prematurely end. The condition for the plateaux is for the algorithm to have reached the same fitness value consecutively for several iterations equal 10% of the remaining iterations during the run. This can be updated if the fitness changes before the condition are met.

5   Experimental Procedure

This section describes the important details pertinent to the simulations. It will cover the creation of solutions via heuristics, some derived constructive heuristics, and other details like assessment metrics and parameters.

5.1   Datasets

In order to maintain continuity in MSSP research, this paper considers the MSSP datasets generated in prior research here [12]. For each class, five problems will be generated for a total of twenty instances. Each of the instances within each class have similar objective values in the sense that all problems of the same class should have optimal values that are very similar. Across problem classes, however, is where larger differences in the objective values will be noted. The description of the problem classes is given in Table 1:

Table 1. A table showing the four problem classes

Class	1	2	3	4
Scenes (S)	10	25	50	100
Actors (A)	5	10	30	60
Locations (L)	3	5	10	15

5.2   Low-Level Heuristic Components

As there are no extant examples of generation constructive hyper-heuristics for the movie scene scheduling problem (MSSP), terminal components had to be created for this work. The ones created are chosen to reflect basic characteristics of the problem state during the construction process and they are described below:

- d: Duration of a given scene.
- a: average wage of the actors attached to a given scene.
- lp: number of already scheduled scenes that share a location with the current scene.

- tc: one divided by the transfer cost from the last scheduled scene to the current given scene. Returns 1 if there are no scheduled scenes.
- td: one divided by the transfer time from the last scheduled scene to the current given scene. Returns 1 if there are no scheduled scenes.
- an: number of actors assigned to the current scene.

The functional components are as follows:

- +: Addition of two inputs.
- −: Subtraction of two inputs.
- *: Multiplication of two inputs.
- /: Protected division of two inputs.
- A: Absolute value of a single input.

5.3   Solution Construction Process

Algorithm 2: MSSP Construction Method

Input: S a heuristic, $scenes$ a set of scenes
Result: sol a constructed solution
1 $sol = \emptyset$;
2 $scores = [\,]$;
3 **while** $scenes \neq \emptyset$ **do**
4 $\quad$ $scores=[scenes.\text{size}()]$;
5 $\quad$ **for** $i < scenes.size()$ **do**
6 $\quad\quad$ $scores[i]=\text{evaluateHeuristic}(S,scenes[i])$;
7 $\quad$ $choice = max(scores)$;
8 $\quad$ $sol+=scenes.\text{remove}(choice)$;

Algorithm 2 describes the process by which a given MSSP solution is constructed. The given scenes need only be added into a vector to form a permutation. The score represents the desirability of adding a given scene into the solution as the next scene to be scheduled. The process uses the maximum score, Line 7, due to the choice of the terminal components.

5.4   Constructive Heuristics

Given that there is a paucity of construction heuristics available for the problem, some construction heuristics would have to be defined and presented here for comparison against the HACO algorithm. These heuristics were manually derived through the study of the problem domain [12].

What follows is a list of the constructive heuristics where c_i denotes constructive heuristic i:

- H_0: The next scene is randomly chosen.
- H_1: The scene with the most actors is chosen.
- H_2: The scene with the fewest actors is chosen.
- H_3: The scene with the longest duration is chosen.
- H_4: The scene with the shortest duration is chosen.

- H_5: The scene with the smallest transfer cost from the prior scene is chosen.
- H_6: The scene with the largest transfer cost from the prior scene is chosen.
- H_7: A scene is chosen randomly from a list of scenes that share the same location as the prior scene. If no such scenes exist, the next scene is chosen randomly.

5.5   Algorithm Parameters

For the HACO algorithm, there are two parameters necessary for its function. These are α and p. The former determines the pheromone desirability and the latter indicates the rate of evaporation. Both parameters are adjusted via schedules to minimise the time required for extensive parameter tuning.

The initial value of p is 0.1 and it will be linearly increased to the value of 0.9. The reason for this is that this achieves an evaporation rate that most promotes an exploration to exploitative trend. When the rate of evaporation is low, the pheromone map will become saturated with pheromone. Over time, the rate of evaporation will increase, and apply a filtering effect, removing all but the best concentrations of pheromone. Thus the algorithm should initially favour exploration before shifting to exploitation.

The parameter α uses to determine its value. It starts from 0.1 before gradually moving to 0.9. The desirability of the pheromone to the ant is α and the desirability of the heuristic to the ant is $(1 - \alpha)$. Thus initially the heuristic will be favoured before a gradual shift to the value of the pheromone. This ensures that initially, the ant will favour novel solutions as the heuristic biases towards very diverse solutions before gradually transitioning the exploitation through favouring the best solutions.

Thus the combination of the two schedules helps facilitate good performance without requiring more fine parameter tuning.

5.6   Experiment Parameters

The experiment conducted for this research consists of simulations where the HACO algorithm and the presented constructive heuristics are applied to the MSSP benchmark instances. The parameters for these experiments are fixed. That is, in terms of the number of runs, the number of iterations, the number of ants n_k considered, and the path limit, p_l. These are:

- n_k: 50
- Number of Runs: 10
- Number of Iterations: 1000
- p_l: 10

With this configuration, the experiments have a computational budget of 50000 fitness evaluations. That is, there can be at most 50000 calls to the fitness function before the algorithm will terminate. These values were decided upon based on the average problem size of the 4 MSSP problem classes to strike a compromise between runtimes and needed computational effort for the larger problems. Every run is independent to remove issues with the stochastic nature of the underlying algorithm.

5.7   Assessment Metrics

In terms of assessment, comparisons will be drawn between the performance (measured in solution quality) between the HACO algorithm and the comparison constructive heuristics. These comparisons will focus on the average performance of the methods over their given runs and standard deviations will be included for the HACO algorithm only as the majority of the constructive heuristics are deterministic in nature. As the instances should have similar objective values within a problem class, the results are aggregated (in terms of the solution quality) across all instances of a given class. This should give the general performance of the technique for problems of that given class. In addition, statistical testing will be performed with the Mann-Whitney U Test [7].

5.8   Technical Specifications

For this research, a computing cluster provided by the University of Pretoria was used. The technical specifications of this cluster are 377 GB RAM, 56 cores at 2.40GhX (Intel Xeon CPU E6-2680 v4), and 1TB of Ceph Storage.

6   Results and Discussion

In this section, the results of the experiments are presented based on the problem classes. In addition, the runtimes and standard deviations of the HACO algorithm (for a single run in minutes) are presented alongside the results. The best results are indicated in bold.

In terms of the results present in Table 2, the HACO algorithm emerges as the best performing technique across all 4 problem classes. As the classes increase in complexity, the gap between the HACO algorithm and the constructive heuristics increases significantly. Of the given 8 constructive heuristics, only H7 can provide solutions that can come somewhat close to the performance of the HACO algorithm. Although not as good as the hyper-heuristic it is still much better for all classes where the other heuristics struggle. The run times and standard deviation of the HACO algorithm also increase as the problem class scales in complexity but this is to be expected for the more complicated problems.

Table 2. Comparison between Constructive Heuristics and HACO Algorithm

Instance	C_S_0	C_S_1	C_S_2	C_S_3
H0	32146.74	117985.58	899236.44	3547519.58
H1	31020.80	110348.20	828723.60	3294085.80
H2	30440.00	111269.60	839935.00	3318200.80
H3	32417.00	118182.00	896614.40	3555482.00
H4	31483.20	115571.40	898698.00	3536160.20
H5	29188.78	111026.86	876850.18	3493635.22
H6	33846.88	127793.08	911692.60	3570439.46
H7	27740.78	103327.82	713886.48	2828560.46
HACO	**23755.20**	**84970.60**	**615750.38**	**2477407.08**
HACO Std Dev	223.58	750.88	5431.46	36257.95
Time Per Run (Min)	0.002	0.0100	0.0400	0.1800

6.1   Statistical Testing

The statistical test is formulated as follows:

– The Null Hypothesis (H0): The distributions of each population are identical
– Alternative Hypothesis (H1): The distributions are not identical.

The tests conducted will be two tailed tests at a 0.05 level of significance. Each test is performed by comparing the normalised results of the HACO algorithm against each of the heuristics.

Table 3. Mann-Whitney U-Test Results

MSSP	H0	H1	H2	H3	H4	H5	H6	H7
U Value	0	0	0	0	0	0	0	0
Z Score	−5.39649	−5.39649	−5.39649	−5.39649	−5.39649	−5.39649	−5.39649	−5.39649
P Value	1E−06	1E−06	1E−06	1E−06	1E−06	1E−06	1E−06	1E−06
Outcome	Reject H0	Reject H0	Reject H0	Reject H0	Reject H0	Reject H0	Reject H0	Reject H0

Based on Table 3, the significance testing reveals that the results of the HACO algorithm are statistically significant in all cases of comparison against the existing heuristics. This, in conjunction with the performance results strongly indicate that the HACO algorithm performed significantly better than the existing heuristics in all problem instances across all problem classes.

6.2   Heuristic Comparison and Interpretation

It is also important to examine the heuristics that are produced by the HACO algorithm to help demonstrate why it was able to outperform the existing heuristics. To this end, four heuristics are presented, one from each of the problem classes (C_S_0 – C_S_3). The heuristic chosen is the best performing heuristic and is presented as constructed.

```
C_S_0:  [*:{*:a,{/:{*:d,{+:td,td}}|lp}}|tc]
C_S_1:  [+:tc,{+:td,{/:tc,an}}]
C_S_2:  [+:td,{*:{/:a,{/:an,tc}}|td}]
C_S_3:  [*:td,{/:an,{*:{*:{A:{A:{+:{-:d,lp}|lp}}}|{A:{/:lp,an}}}|d}}]
```

The HACO algorithm generates heuristics in a process that follows a depth-first strategy. When considering a functional component, the algorithm will fully specify its first input before moving to the second. So if there are going to be nested or compositional elements, these will tend to occur towards the first operand in a given operator. The heuristics presented above represent the best heuristics created for various problem classes and are chosen to help determine some trends in the way the algorithm produces heuristics for a variety of different problem conditions.

For the presented heuristics there is a trend towards considering some combination of factors. Specifically the terminal components: transfer costs and transfer times, td and tc.

The least represented terminal component, in terms of the examples presented above, would be the wages of the actors as a is one of the least represented components which suggests that the wages of the actors are an insignificant factor when deciding how to schedule a scene. This does make sense considering that the actor wages are a relatively fixed value. The exact daily wage of each actor is fixed and their production cost depends more on the number of scenes they are in and their proximity to other scenes than their strict value.

Another pertinent fact is the prevalence of the lp terminal component in the presented heuristics above, especially the heuristic for C_S_3 where lp is featured thrice. This component quantifies the number of scenes that share the location of the current scene. This would likely help to explain the performance of the algorithm for the larger problem. As problems become more complicated, it would suggest that scene proximity becomes the most deciding factor in optimising a solution. It is also apt to note that for the largest problem class, the heuristic generated was the largest although the second-largest class did not. This might suggest that the largest problem class might differ more substantially than its prior class.

These presented heuristics demonstrate the algorithm can produce heuristics (of varying lengths) that combine a variety of factors (but not necessarily all) when attempting to produce a heuristic. The existing heuristics tend to focus on a single variable or characteristic of the problem and build their solutions around it. A good generation hyper-heuristic can produce a heuristic that combines multiple influences in a generalised way. In particular, constructive heuristics cannot typically, by themselves, produce optimal solutions. The are intended for providing good initial starting points. So the value of a good generative constructive hyper-heuristic is in generating construction heuristics that produce better results than existing heuristics.

In this way, ideas derived from existing heuristics can be incorporated into generated heuristics without specifying beforehand how this combination should be produced. Of course, the value of the constructed heuristics will heavily depend on the value of the provided components. Better components will naturally increase the potential of the generated heuristics.

As a problem, the MSSP relies heavily on the location assignments of the scenes. While the wages of actors are one component of the problem cost, scene locations have a multifaceted influence on the cost of solutions. Where two scenes might be scheduled dictates the times to move between them, for both production and the crew, which plays the biggest role in wage costs as well. H7 works by assigning scenes based on their proximity (location-wise) to other scenes and intuitively, this strategy makes a lot of sense. If all the scenes of a given location are shot in a batch, this means travel times will only be incurred when all of those scenes are finished. Additionally, if many actors overlap their scenes in this batch of scenes, they will minimise downtime while waiting and their travel times. These factors probably contribute the most to the success of H7.

With that being said, the success of the HACO algorithm then is heavily influenced by the ability of the algorithm to incorporate this kind of information into the heuristics it generates. Accounting for some combination of factors like scene proximity and actor density allows the HACO algorithm to generate heuristics that combine multiple considerations into a heuristic that outperforms the available human-derived heuristics by a wide margin. That is not to say that the HACO algorithm is perfect. The MSSP domain is a new field and additional work will need to be done to develop it in terms of its low-level heuristics.

There remains the potential that new heuristics could be developed that prove better suited for the task. However, what has been soundly demonstrated is the ability of an ant-based generation constructive hyper-heuristic to generate competitive heuristics for a wide range of instances of the MSSP. It is capable of doing this by allowing for the combination of multiple characteristics from the problem to be included in its generated heuristic.

7   Conclusion

This paper proposed the application of the HACO algorithm to the task of generating new generation constructive hyper-heuristics for the MSSP, with the main aim being to automate construction heuristic creation in a previously unexplored domain. The results of the experimental trials demonstrated that the HACO algorithm was successfully able to produce new constructive heuristics that could be applied to create solutions of better quality than a wide arrangement of existing human-derived heuristics. The ability of the algorithm to, with relatively few components, incorporate a wide array of information into its heuristics is the primary reason it performed so well, although the relative infancy of the MSSP means that the opportunity for better human-derived heuristics remains significant. From this research, several potential avenues for future exploration exist. In particular, the extension of the HACO algorithm method towards generating new perturbative heuristics for the problem as well as examining how a hybridised hyper-heuristic could perform in this domain in contrast to the singular hyper-heuristic utilised here.

Acknowledgments. This work was funded as part of the Multichoice Research Chair in Machine Learning at the University of Pretoria, South Africa. This work is based on the research supported wholly/in part by the National Research Foundation of South Africa (Grant Numbers 46712). Opinions expressed and conclusions arrived at, are those of the author and are not necessarily to be attributed to the NRF.

References

1. Aarseth, E.: The culture and business of cross-media productions. Pop. Commun. **4**(3), 203–211 (2006). https://doi.org/10.1207/s15405710pc0403_4
2. Garcia de la Banda, M., Stuckey, P., Chu, G.: Solving talent scheduling with dynamic programming. INFORMS J. Comput. **23**, 120–137 (2011). https://doi.org/10.1287/ijoc.1090.0378
3. Cheng, T.C.E., Diamond, J., Lin, B.: Optimal scheduling in film production to minimize talent hold cost. J. Optim. Theory Appl. **79**, 479–492 (1993). https://doi.org/10.1007/BF00940554
4. Gregory, P., Miller, A., Prosser, P.: Solving the rehearsal problem with planning and with model checking. Eur. Conf. Artif. Intell. **16**, 157–171 (2004)
5. Liu, Y., Sun, Q., Zhang, X., Wu, Y.: Research on the scheduling problem of movie scenes. Discret. Dyn. Nat. Soc. **2019**, 1–8 (2019). Article ID 3737105
6. Long, X., Jinxing, Z.: Scheduling problem of movie scenes based on three meta-heuristic algorithms. IEEE Access **PP**, 59091–59099 (2020)
7. Mann, H.B., Whitney, D.R.: On a test of whether one of two random variables is stochastically larger than the other. Ann. Math. Stat. **18**(1), 50–60 (1947)
8. Nordstrom, A., Tufekci, S.: A genetic algorithm for the talent scheduling problem. Comput. Ind. Eng. **21**, 927–940 (1994)
9. Sakulsom, N., Tharmmaphornphilas, W.: Scheduling a music rehearsal problem with unequal music piece length. Comput. Ind. Eng. **70**, 20–30 (2014)
10. Singh, E., Pillay, N.: A comparison of ant-based pheromone spaces for generation constructive hyper-heuristics. Under Review (2021)
11. Singh, E., Pillay, N.: Haco: Supplementary material (2021). https://drive.google.com/file/d/1DHtMBkwTfT-h4sEIYsqa7-Nx9rD31sMZ/view?usp=sharing
12. Singh, E., Pillay, N.: Ant-based hyper-heuristics for the movie scene scheduling problem. In: Proceedings 2021 ICAISC International Conference for Artificial Intelligence and Soft Computing (To appear June 2021)

Author Index

Printed in the United States
by Baker & Taylor Publisher Services